SCIENCE IN
POPULAR CULTURE

SCIENCE IN POPULAR CULTURE

A Reference Guide

A. Bowdoin Van Riper

GREENWOOD PRESS
Westport, Connecticut • London

Library of Congress Cataloging-in-Publication Data

Van Riper, A. Bowdoin.
   Science in popular culture : a reference guide / A. Bowdoin Van Riper.
    p. cm.
   Includes bibliographical references and index.
   ISBN 0–313–31822–0 (alk. paper)
   1. Science in popular culture. I. Title.
  Q172.5.P65V36 2002
   306.4'5—dc21   2001055616

British Library Cataloguing in Publication Data is available.

Library of Congress Catalog Card Number: 2001055616
ISBN: 0–313–31822–0

First published in 2002

Greenwood Press, 88 Post Road West, Westport, CT 06881
An imprint of Greenwood Publishing Group, Inc.
www.greenwood.com

Printed in the United States of America

∞™

The paper used in this book complies with the
Permanent Paper Standard issued by the National
Information Standards Organization (Z39.48–1984).

10 9 8 7 6 5 4 3 2 1

to Julie Newell

colleague • wife • inspiration

CONTENTS

ACKNOWLEDGMENTS

This book would not exist if Debra Adams, an acquisitions editor at Greenwood Publishing Group, had not seen a need and set out to fill it. The concept and basic structure of the book are hers, and her astute suggestions during the writing process have shaped countless details of its format, content, and style. We have never met face to face, but our interactions by phone and e-mail have been models of the writer-editor partnership.

Production editor Megan Peckman and her staff have also been a great help, especially in guiding me through the intricacies of choosing illustrations. Copyeditor Pelham Boyer artfully improved my prose in many places and caught several errors of fact that had slipped by me.

Over the year and a half it took to write *Science in Popular Culture*, friends, colleagues, and family members have patiently answered what must have seemed like an endless series of bizarre questions. They have also given me places to write, editorial advice, technical expertise, and a chance to fine-tune my writing style by "test driving" parts of the book as members of its target audience. Thanks, in alphabetical order, to: John Lockhart, Joe Mundt, Julie Newell, Alan Riley, Judy Riley, John Szucs, Jan Van Riper, Tony Van Riper, James Whitenton, and those whose names I have inevitably forgotten. All have helped to make this a better and more complete book, and all have my heartfelt thanks and appreciation.

I also owe three other debts, less concrete but no less significant. The first is to my daughter Katie, now six, who enriched the book by helping me to rediscover the world of children's entertainment. The second is to Jim Berkowitz, pop culture maven *extraordinaire*, who long ago taught me to read what is written below the surface of popular culture. The third is to the dedicated creators of pop culture Web sites, who put the answers to seemingly unanswerable questions ("What's the name of that cartoon where Bugs Bunny is trying to fly the giant plane?") at my fingertips. The Internet may not have transformed the world, but it has surely transformed the study of popular culture.

CONVENTIONS USED IN THIS BOOK

The metric system, though virtually universal in science, is still unfamiliar to most Americans. All measurements in this book are, therefore, given in the English system.

Dates given for creative works reflect first publication for books, plays, and stories; first release for movies; first network run for TV series; and the equivalents for other media. All works are listed by their U.S. titles. Names of book authors are given as they appear on the title page, regardless of the author's real name (e.g., Mark Twain, not Samuel L. Clemens). Popular songs, unless otherwise specified, are attributed to the songwriters rather than to the performers. Names of movie and television characters are followed, in parentheses, by the names of the actors who played them.

All names, titles, dates, and similar information about creative works have been checked against standard reference sources (listed in the General Bibliography) and are believed to be accurate. Any errors and omissions are unintentional.

INTRODUCTION

When a character in a movie or television show easily finds a parking space in front of a downtown building at midday, we accept it as dramatic license. We say "That's Hollywood!" to ourselves or the person next to us and turn our attention back to the story—knowing all the while that in the real world, there is seldom a parking space empty when we need one. The dramatic license is obvious because it involves something that we do every day. We excuse it because we know that watching the hero circling the block would derail the story. Other examples of dramatic license pass unnoticed because they are outside of our everyday experience. Sailors wince when the skipper of a fictional schooner sends crew members aloft to furl the mainsail. Gun enthusiasts groan when a fictional villain pulls out a silenced revolver. Paleontologists chuckle when the cover of *Jurassic Park* features a dinosaur that lived not in the Jurassic but in the later Cretaceous period. Audience members who are experts in other areas enjoy the show, untroubled.

This book is an attempt to separate reality from dramatic license in popular culture's treatment of science and of some of the technologies deeply influenced by it. Each of its eighty-one entries deals with a science-related object, idea, person, process, or concept. Each briefly summarizes the current understanding of the topic, then discusses its portrayal in popular culture and, where possible, the roots of that portrayal. The titles of the entries sometimes, for the sake of clarity, reflect popular culture rather than science: No compact, scientifically accurate phrase covers the same ground as "death ray" or "miracle drug." Each entry concludes with a list of related entries and a brief list of suggested readings that (in the interest of accessibility) emphasizes books, large-circulation periodicals, and established Web sites. The bibliography at the end of the book covers general works on science, popular culture, and science *in* popular culture.

This book is designed to serve multiple purposes. One is to separate fact from fiction in popular culture's depiction of particular scientific and technological topics. A second is to identify exemplary treatments of particular scientific topics in popular culture. A third is to explore recurring patterns in popular culture's depictions of science and technology in general. Individual entries may also serve as brief introductions to, and guides to further reading about, their subjects.

"Science" is both a body of knowledge and the process used to expand and revise it. The body of knowledge includes discrete facts, patterns that order them, and explanations of why those patterns exist. The process of expanding and revising that body of knowledge has many elements, among them observation, experimentation, mathematical analysis, and computer modeling. All can be used to test new explanations and reexamine old ones. The results of the process are shaped, but not determined, by the cultural context in which it takes place: influences such as political tensions, economic pressures, religious beliefs, personal ambitions, and institutional rivalries.

Science's multifaceted nature complicates the process of defining its boundaries. Where those boundaries fall—which ideas they include and which they exclude—has been the subject of debate for centuries. For practical reasons such as length, this work defines "science" conservatively and draws its outer boundaries narrowly. The majority of the entries deal with topics from the familiar "natural science" disciplines: chemistry, physics, biology, geology, astronomy, meteorology, and biological anthropology. Some entries, however, cross that boundary, in order to deal with topics in medicine (epidemics), psychology (dreams, intelligence), and especially technology (computers, robots, space travel). A handful of essays focus on subjects that fall well outside mainstream science. Some (UFOs, psychic powers) treat ideas championed by small groups of enthusiasts but rejected or viewed with intense skepticism by mainstream scientists. Others (giant insects, time travel, matter transmission) deal with things that current understandings of nature suggest are impossible. I have included these boundary-crossing topics because, in the world of popular culture, they are emphatically part of "science."

"Popular culture" is easy to define in general terms but hard to define precisely. Its overlap with "folk culture" and "mass culture," substantial but incomplete, is one barrier to a precise definition. Its uncertain place on the spectrum ranging from "low" to "high" culture is another. These distinctions are even less clear, and the concept of "popular culture" even more problematic, in centuries before the twentieth. All these issues are significant and deserve close consideration—but not in the context of a book like this one. Popular culture, for the purposes of this book, includes any creative work designed to appeal to a large audience. It includes movies, television programs, and popular music, along with more ephemeral material like printed cartoons, advertisements, commercial illustrations, and jokes. It also includes most of the stock—all fiction and most nonfiction—in an average chain bookstore.

Even in a book focused, like this one, on the United States since 1900, this definition creates an enormous pool of works to draw examples from. This book is designed to cover the widest possible range of media, genres, and decades and to emphasize famous, readily accessible works over rare and obscure ones. The omission of a particular work from a particular entry should not, therefore, be interpreted as a judgment of its artistic value.

These entries are not intended to be the last word on their subjects. Scientific discoveries made after this book goes to press will reinforce some of its claims and undercut others. Creative works that appear after it will do the same. It is the nature both of science and of popular culture to be fluid. That fluidity keeps them fresh, but it means that generalizations about them need to be read with the passage of time firmly in mind.

SCIENCE IN
POPULAR CULTURE

Acceleration

Acceleration, in everyday usage, is an increase in speed; scientists use the term more broadly to mean *any* change in motion. An object accelerates, in this broader sense of the word, when it speeds up, slows down, or changes direction. Objects accelerate only when a force is applied to them. How *much* force will produce how *much* acceleration depends on the mass of the object—that is, on how much "stuff" it contains. A force capable of accelerating a rifle bullet from rest to the speed of sound would hardly budge a cannonball. The equation that describes this intuitive idea that "force equals mass times acceleration" ($F = ma$) was the second of the three laws of motion devised by Isaac Newton in the mid-seventeenth century. Using the equation, scientists (or anyone else who might want to know) can calculate how much force (F) must be applied to an object of known mass (m) to produce a desired acceleration (a).

Most of popular culture tries, most of the time, to depict the world of everyday material objects realistically. The physical laws governing acceleration are thus shown working more or less as they do in the real world. Tales of fast cars, for example, often depend on audiences' grasp of the relationship between force, mass, and acceleration. The singer in the Beach Boys' "Shut Down" (released on *Surfin' USA,* 1963) wins a drag race because his Corvette has a better force-to-mass ratio, and thus better acceleration, than the other car. Key moments in movies like *The Blues Brothers* (1980) and *The Road Warrior* (1982) hinge on the fact that the hero's car (despite its decrepit appearance) has a better force-to-mass ratio than the villain's. "Our Lady of Blessed Acceleration, don't fail me now!" intones Elwood Blues (Dan Aykroyd), preparing to jump across an open drawbridge in his decommissioned police car.

When popular culture aims for the fantastic rather than the realistic, however, these limitations disappear. Large accelerations no longer require large forces. Large objects require no more force to accelerate than small ones. Acceleration becomes, in effect, something that can be arbitrarily applied to or removed from any object at any dramatically convenient time.

Fictional characters routinely take advantage of these loopholes by accelerating massive vehicles to enormous speed in seconds. The imperial Death Star featured in *Star Wars* (1977) is the size of a small moon yet can move between star systems in a matter of days or even hours. The hapless characters of the TV series *Space: 1999* (1975–1977) visit a different planet nearly every week while riding an *actual* moon—Earth's, which is blown out of its orbit in the pilot episode. The force necessary to achieve such accelerations is staggering—*if* Newton's Second Law is in effect. Cartoon characters also benefit from the apparent suspension of the Second Law. Fred Flintstone can accelerate a car made of stone and logs with his bare feet. Bugs Bunny, in "Baseball Bugs" (1946), launches a blazing fastball from the pitcher's mound, then streaks past it and arrives at home plate in time to catch it.

The suspension of the Second Law also allows massive, fast-moving objects to stop and change direction virtually at will. The tall-tale ballad "The Legend," sung by Jerry Reed in the film *Smokey and the Bandit* (1977), describes how one of the film's heroes stopped a runaway eighteen-wheeler by dragging his feet. Bugs Bunny brings a nose-diving airplane to a dead stop only a few feet from the ground in "Falling Hare" (1943), laughingly saying "Lucky for me this thing had air brakes." An astronaut in the film *Mission to Mars* (2000) accelerates steadily toward a distant goal and then, realizing it is unreachable, reverses direction with a single brief burst of propellant. Cartoon characters, both biological and mechanical, make virtually right-angle turns at high speed without difficulty. So does Han Solo's *Millennium Falcon*, as it flies through an asteroid field in *Star Wars: The Empire Strikes Back* (1980) and through a newly constructed Death Star in *Star Wars: Return of the Jedi* (1983).

Observers who interpret UFOs as spacecraft from other worlds often, significantly, support their views by noting to the UFOs' apparent ability to make rapid speed changes and right-angle turns. Nothing known to or built by humans, they argue, could accelerate like that.

Related Entries: Inertia; Newton, Isaac; Space Travel, Interplanetary; Space Travel, Interstellar

FURTHER READING

Asimov, Isaac. *Motion, Sound and Heat*. New American Library, 1969.

Krauss, Lawrence M. *The Physics of Star Trek*. Basic Books, 1995. Chap. 1. Discusses the application of Newton's laws to space flight.

March, Robert H. *Physics for Poets*. 4th ed. McGraw Hill, 1995. Non-technical explanations of the laws of motion.

Action and Reaction, Law of

The third of Isaac Newton's three laws of motion states: "For every action, there is an equal and opposite reaction." It is the most familiar of the three laws but also the farthest removed from the realm of common sense. The idea that a chair exerts an upward force on its occupant equal to the downward force the occupant exerts on the chair is, for most people, deeply counterintuitive.

Intuitive or not, the law has significant, wide-ranging effects. Jet and rocket engines work because the force that drives hot gasses out the back end is matched by a force of equal intensity that drives the engine (and anything attached to it) in the opposite direction. A gun recoils when fired because the force that drives the bullet down the barrel toward the target is matched by a force of equal intensity that drives the barrel away from the target. Astronauts working in zero gravity must securely anchor their feet in order to keep the forces they apply to their tools from also pushing them away from the work.

The law of action and reaction, because it governs motion, is seldom visible except in moving images. Even the branches of popular culture that are built on moving images—movies and TV—seldom place it in the foreground. *Destination Moon* (1950), the first realistic depiction of space travel on film, has the builders of the world's first spaceship explain it in detail in order to convince skeptical backers (and audience members) that a rocket will work in space, although "there's nothing to push on." Since *2001: A Space Odyssey* (1968), however, most realistic treatments of space travel have treated the law (like the zero-gravity environment that makes it significant) as part of the background. Astronauts in *Apollo 13* (1996), *Deep Impact* (1998), and *Space Cowboys* (2000) anchor themselves to their work surfaces but don't stop to explain why.

Far more often, the law of action and reaction is conspicuous in popular culture by its absence. Like other laws of physics, it is routinely suspended in the fictional universes of movies and TV programs in order to satisfy dramatic conventions. The closing scene of the James Bond thriller *Moonraker* (1979) has Bond (Roger Moore) and his beautiful colleague Dr. Holly Goodhead (Lois Chiles) making love in zero-gravity aboard an orbiting space station. Bond films have, since 1962, nearly always ended with such a scene. Tradition was, for the filmmakers, reason enough to ignore the substantial complications that the law of action and reaction would create for weightless lovers.

Suspensions of the law of action and reaction are most common, however, in scenes involving firearms. The recoil that snaps a pistol shooter's hand up and back, or bruises a careless hunter's shoulder, is a direct product of Newton's Third Law. The bigger and more powerful the shell being fired, the greater the recoil involved. In movies and TV, however, dramatic convention routinely trumps the laws of nature. Handheld weapons, no matter how large, produce no more recoil on screen than a .22-caliber target pistol. The hero can fire, without even flinching, a bullet capable of blowing the villain several feet straight backward.

This convention allows action heroes like John Rambo (Sylvester Stallone) or Col. James Braddock (Chuck Norris, in the *Missing in Action* series) to fire from the hip machine guns that would normally be attached to something more substantial—like a truck. It allows them to do so, moreover, without suffering broken bones, strained joints, or even visible bruises from the effect of the gun slamming against their bodies hundreds of times a minute. Indeed, it allows them to fire continuously without expending any effort to keep the gun barrel pointed at the target. One of the rare plausible exceptions to this is Arnold Schwarzenegger's character from the *Terminator* films (1984, 1989), but he has the considerable advantage of being a robot.

The second part of the convention, which allows "Dirty Harry" Callahan (Clint Eastwood) to blow felons through plate-glass windows without losing his trademark sneer, involves a more subtle defiance of reality. The law of action and reaction insists that a bullet capable of physically *knocking over* the man it hits must have been propelled by enough force to also knock over the man who fired it—even if he is the good guy. It will be a great day for science, if not for box office receipts, the first time Hollywood shows a hero firing an enormous weapon and, due to Newton's Third Law, landing unceremoniously on his backside.

Related Entries: Acceleration; Gravity; Inertia

FURTHER READING

Asimov, Isaac. *Motion, Sound and Heat*. New American Library, 1969.

March, Robert H. *Physics for Poets*. 4th ed. McGraw-Hill, 1995. Nontechnical explanations of the laws of motion.

Newton, Michael. *Armed and Dangerous*. Writer's Digest, 1990. A fiction writer's guide to firearms; critiques dramatic conventions that defy the laws of physics.

Alternate Worlds

We commonly distinguish between the "real world" we inhabit and the imagined worlds of popular culture. Most imagined worlds are realistic portraits of the real world, peopled with fictional characters and subtly "improved" by dramatic license. Imagined worlds distinct from the real world are the province of science fiction and fantasy: the exotic planets of *Star Wars*, for example, and the magical realms of Wonderland, Oz, or Middle Earth. "Alternate worlds," apparently similar to the real world but different in significant ways, are an intermediate category. They come in two types, one rooted in evolutionary theory and the other rooted in quantum mechanics.

Earth and the community of living things inhabiting it are complex, interdependent, continually evolving systems. The condition of either system depends, at any given moment, not only on its own prior condition but also on the current condition of the other. The evolution of the earth and its living inhabitants does not, therefore, follow a predetermined course, and the evolution of humans and the earth as we know it was not inevitable. Earth is well suited to support life, scientists note, but life need not have taken the specific forms familiar to us. Roll the evolutionary dice again, starting with the same conditions, and the end result might be very different.

Stories set in the first type of alternate world assume that at some crucial moment in the past the evolutionary (or historical) dice *did* fall differently. Characters in these stories inhabit (and take for granted) the very different "real world" that developed in place of ours. Occasionally the differences are biological. Harry Harrison's "Eden" trilogy of novels (1984–1988), for example, is set on an Earth where the dinosaurs did not become extinct 65 million years ago. Harry Turtledove's novel *A Different Flesh* (1988) imagines that *Homo erectus* survived in the New World while *Homo sapiens* evolved in the Old. More often, the differences are historical. Brendan DuBois's novel *Resurrection Day* (2000), for example, is set in 1972—ten years after the Cuban missile crisis erupted into full-scale nuclear war that destroyed the Soviet Union and crippled the United States. Our reality, in which war was averted, is only a might-have-been pipe dream for DuBois's characters.

Quantum mechanics is a branch of physics concerned with the behavior of particles smaller than atoms. Its "many-worlds interpretation" accounts for certain quirks in that behavior by proposing that every event with multiple possible outcomes causes the world (meaning, in everyday usage, "the universe") to split into multiple worlds, identical at the moment of splitting except that in each one a different possible outcome is played out. These multiple worlds are independent of each other. Detection of other worlds may be theoretically possible, but communication between them is not. The many-worlds interpretation implies that at the macroscopic level our "real world" is only one in a nearly infinite collection of parallel universes, each of which is equally real to *its* inhabitants, and each of which differs in varying degrees from ours.

The second type of alternate-world story assumes that parallel universes exist and that travel between them *is* possible. The means of travel matters less than the result: inhabitants of one universe find themselves in a different universe that is partly (but never entirely) like their own. Several episodes of *Star Trek* (1966–1969) and *Star Trek: Deep Space Nine* (1993–2000), beginning with 1967's "Mirror, Mirror," plunge the lead characters into a parallel universe ruled by violence and ruthlessness. The TV series *Sliders* follows its four heroes through a different parallel universe each week as they try to get home to their own. Another quartet of heroes goes universe hopping on purpose in Robert A. Heinlein's novel *The Number of the Beast . . .* (1982), and a harried New Yorker escapes into a *slightly* different version of his world in Jack Finney's novel *The Woodrow Wilson Dime* (1968). The "holodeck," a form of virtual reality technology featured in the *Star Trek* saga since 1987, allows users (in effect) to design and enter their own parallel universes for recreation.

Ideas such as the many-worlds interpretation of quantum mechanics and the unpredictability of evolution suggest that we are not as special as we like to think. Stories of alternate worlds, though rooted in those ideas, promote the opposite view. The alternate realities they depict are never as attractive as ours. We live, they imply, in "the best of all possible worlds."

Related Entries: Evolution; Evolution, Human; Time Travel

FURTHER READING AND SOURCES CONSULTED

Gould, Stephen Jay. *Full House*. Harmony Books, 1996. Treats contingency in evolution; source of the discussion in this essay.

Jones, Douglas. "The Many-Worlds Interpretation of Quantum Mechanics." 4 May 2001. 5 June 2001. <http://www.station1.net/DouglasJones/many.

htm>. A brief, clear, nonspecialist's explanation; source of the discussion in this essay; links to more detailed sources.

Schmunck, Robert B. *Uchronia: The Alternate History List*. 11 April 2001. 11 June 2001. <http://www.uchronia.net>. An exhaustive, searchable bibliography of alternate-world stories, with a masterful introduction.

Androids

Androids are robots designed to look and act human. They exist, now and for the foreseeable future, only in fiction. "Animatronic" human figures like those used in Disney World's "Pirates of the Caribbean" and "Hall of Presidents" attractions are androids, but only in the broadest sense. Unable to sense or respond to the world around them, they move only in preprogrammed ways and speak only prerecorded words. True androids would—like the humans they simulate—have full, fluid mobility in both body and limbs. They would be intelligent enough to interact, in flexible and adaptable ways, with humans, other androids, and the material world. They would be able to interpret casual human speech accurately and to produce a reasonable facsimile of it themselves.

A robot that achieved even one of these goals would be a technological step far beyond the current state of the art. A closer look at the problem of mobility shows why. Mobile robots have traditionally been designed with wheels, to run on flat surfaces such as warehouse and factory floors. More sophisticated robots, like the Mars explorer *Sojourner*, can traverse rough terrain but still use a carlike design: wheels and a low center of gravity. A true android, however, would carry itself like a human rather than a car: vertically, with its center of gravity three feet or more above the relatively small base provided by the soles of its two feet. Motions such as bending, reaching, or lifting would alter the center of gravity and unbalance the android. The android's brain would, therefore, constantly have to evaluate and compensate for these motions, all while focusing on the task that made them necessary. Walking, with its constant shifting of weight and attitude, would require even more complex adjustments. Stair climbing, the most challenging form of everyday human walking, would be a nightmare for android designers.

The technological problems of making a humanoid robot with fluid, humanlike mobility are probably soluble. The resulting machine, however, is likely to be extremely complex, high-maintenance, and expensive. Would-be builders and marketers of commercial androids would face a difficult question from prospective customers: why use an android at all? What would an android offer, aside from novelty, that would justify its cost? What could an android do that a human or conventional (non-humanoid) robot could not do as well or better, and for less money?

The androids portrayed in popular culture easily meet the technological challenges and quietly sidestep the economic uncertainties that would bedevil the real thing. They move, think, and speak as fluidly as flesh-and-blood humans, and they are so reliable that mechanical failures rarely disrupt the illusion that they *are* human. The illusion is so perfect, in fact, that fictional androids routinely do well in jobs that challenge flesh-and-blood humans. Commander Data (Brent Spiner), of TV's *Star Trek: The Next Generation* (1987–1994), is third in command of a giant starship. R. Daneel Olivaw, of Isaac Asimov's novels *The Caves of Steel* (1954) and *The Naked Sun* (1957), is a talented detective. Zhora (Joanna Cassidy), one of the fugitive androids in the film *Blade Runner* (1982), has a brief but apparently successful career as an exotic dancer. The list of examples could be much longer: androids as soldiers, prostitutes, assassins, interpreters, executive secretaries, spaceship pilots, theme-park actors, household servants, and suburban housewives. Their ability to do these jobs is never in doubt; the androids are "human" enough to step easily into human roles in society.

It is precisely that paradox—characters that are seemingly human, yet also nonhuman—that drives most stories about androids. Rick Deckard (Harrison Ford), the emotionally barren hero of *Blade Runner*, discovers that the androids he was hired to hunt down and kill are more "human" than he is. Over seven seasons of *Star Trek: The Next Generation*, Data seeks to understand human emotions so that he can experience them for himself. The android heroes of the films *Bicentennial Man* (2000) and *A.I.* (2001) seek, Pinocchio-like, to become human. The women in Ira Levin's novel *The Stepford Wives* (1972) move in the opposite direction: their husbands quietly replace them with docile, compliant android lookalikes. Visitors to an adult amusement park in *Westworld* (1973) happily act out their fantasies of casual violence and commitment-free sex with androids who look and act "just like the real thing"—until the androids rebel against such treatment.

Popular culture has good dramatic reasons to take the technological sophistication and everyday utility of androids for granted. The stories it tells about androids aren't *about* androids, in the sense that *Jurassic Park* isn't about dinosaurs, so much as it is about genetic engineering. They are really stories about humans and what it means to be one.

Related Entries: Cloning; Cyborgs; Intelligence, Artificial; Robots

FURTHER READING

Schelde, Per. *Androids, Humanoids, and other Sci-Fi Monsters: Science and Soul in Science Fiction Films.* New York University Press, 1994.

Telotte, J.P. *Replications: A Robotic History of the Science Fiction Film.* University of Illinois Press, 1995. Discussions of androids as vehicles for exploring the nature of humanity.

Willis, Chris. *Android World.* 10 August 2001. Android World, Inc. 15 August 2001. <http://www.androidworld.com/index.htm>. Comprehensive, award-winning site tracking current developments in android technology.

Atomic Energy

Albert Einstein showed early in the twentieth century that matter and energy could be converted into one another. His famous equation $E = mc^2$—in which E stands for energy, m for mass, and c for the speed of light (300,000,000 meters/second)—quantifies this relationship. The magnitude of c makes the relationship wildly unbalanced: the conversion of even a tiny amount of matter produces huge quantities of energy, and producing matter by conversion demands energy on a grand scale. Matter-energy conversions take place when a large atomic nucleus splits into two or more smaller pieces (fission) or when two small nuclei combine into a single larger one (fusion). Both processes usually consume energy and yield products that have slightly more mass than the original ingredients. When certain kinds of nuclei are involved, however, the process consumes small amounts of mass and yields huge quantities of "atomic energy."

Atomic energy has been tapped, by both fission and fusion, for both military and civilian uses. The United States exploded the first fission bombs in 1945, and the first fusion bombs (also known as "hydrogen bombs") after the element used as fuel in 1952. The two major "peaceful" uses of nuclear energy—ship propulsion and electric-power generation—were also pioneered in the mid-1950s. The U.S.S. *Nautilus*, the first ship in the "nuclear navy" envisioned by Adm. Hyman Rickover, was launched in 1953. The first experimental nuclear power plant went on line in 1951, and the first commercial plant began supplying electricity to the city of Pittsburgh in late 1957. A wide range of other proposed uses for nuclear energy—nuclear-powered cars and the use of atomic bombs for large-scale earth-moving projects—died in the planning stages. Nuclear-powered cargo ships and spacecraft achieved limited development but faced nonexistent demand and growing public opposition. Commercial fusion power plants have also failed to materialize: containing fusion reactions (at temperatures exceeding 1,000,000° Celsius) is an enormous technical challenge, and attempts to replicate the room-temperature "cold fusion" announced in 1992 have been fruitless.

Popular culture rarely makes clear distinctions between fission and fusion. Fission and fusion weapons alike are simply "nuclear weapons," and fusion-based power plants receive little attention, presumably because of the difficulty of sustaining fusion except under laboratory conditions. The key distinction is between two conflicting images of atomic energy. The optimistic view portrays it as a powerful-but-pliant servant, the pessimistic view as a barely controllable demon, always on the verge of a rampage.

NUCLEAR WEAPONS

Both fission and fusion bombs are designed to create, at the instant they are detonated, an uncontrolled nuclear reaction capable of producing massive quantities of energy. The nearly instantaneous release of all that energy—in the form of heat, blast, and radiation—is what makes nuclear weapons far more destructive than conventional weapons. The largest conventional bombs routinely used during World War II contained a little less than two tons of high explosive. "Little Boy," the first fission bomb used in war, unleashed the equivalent of 20,000 tons of high explosive on Hiroshima in August 1945. The two-ton conventional bombs had been nicknamed "blockbusters" for their supposed ability to gut an entire city block, rather than just a single building. "Little Boy" and its offspring had the potential to be "city busters," a term sometimes applied to the still-more-powerful fusion bombs of the 1950s.

The optimistic view of nuclear weapons assumes that they can be treated as more powerful (and so more efficient) versions of conventional weapons. It also assumes that—again like conventional weapons—their effects will be limited to the immediate target area. An October 1951 special issue of *Collier's* magazine, describing a hypothetical U.S.-Soviet war, treats nuclear attacks on cities as equivalent to the conventional bombing campaigns of World War II. The nuclear attacks in *Collier's* wreck war industries and crush morale more efficiently, however, because they cause more destruction in less time. *Fail-Safe* (novel 1962, film 1964) treats the nuclear destruction of Moscow and New York City in similar terms: as a great tragedy, but one from which recovery is possible. Terrorists and megalomaniacs threatening to set off nuclear weapons are a standard plot device in action movies from *Goldfinger* (1964) to *Broken Arrow* (1996) and *The Peacemaker* (1997), but the threat is always confined to a single city. When the hero of Michael Crichton's *The Andromeda Strain* (novel 1970, film 1971), races to disarm the nuclear bomb that is about to destroy a secret government laboratory, he is worried less about the explosion than about its effects on a lethal extraterrestrial virus.

The pessimistic view assumes that the effects of nuclear weapons can be neither contained nor predicted. It treats "limited nuclear war" as an oxymoron and assumes that *any* use of nuclear weapons is likely to provoke a full-scale nuclear exchange that will lay waste to the earth. Some stories from this tradition, like Nevil Shute's *On the Beach* (novel 1957, film 1959) and *The Day After* (TV film, 1983) assume that existing nuclear weapons can obliterate human life. Others—films like 1964's *Dr. Strangelove* and 1970's *Beneath the Planet of the Apes*—invent a "doomsday bomb" (conceived, but never built, in the real world) that can do the job in a single explosion. Still others suggest that nuclear explosions may trigger an environmental catastrophe. "Nuclear winter," the planetwide ice age described in Jonathan Schell's nonfiction bestseller *The Fate of the Earth* (1982), is one familiar example of such a catastrophe. So, at the other end of the plausibility spectrum are the irradiated monsters popular in 1950s science-fiction films.

NUCLEAR POWER PLANTS

Nuclear power plants work by controlling the same fission and fusion reactions that take place uncontrolled when a nuclear weapon explodes. The heat generated by the nuclear reaction boils water that in conventional power plants would have been heated by burning coal or oil. The steam can then be used—as in a conventional power plant—to drive a turbine that, in turn, drives a ship's propeller or an electrical generator. Though widely used for both purposes since the late 1950s, nuclear power plants have been the center of intense political controversy. Advocates point to their limited demand for fuel and their lack of air-polluting chemical emissions. Critics emphasize the problem disposing of radioactive waste and the potential loss of life and property that would result from a serious accident.

Stories involving large vehicles with nuclear propulsion take an implicitly optimistic view of the technology. The nuclear submarines in Tom Clancy's novel *The Hunt for Red October* (1985) are—like the real-world versions—so reliable and efficient that the characters take those qualities for granted. The breakdowns that *do* occur, as in Robert Heinlein's short story "The Green Hills of Earth" (1947) or the movie *Star Trek II: The Wrath of Khan* (1982), rarely destroy the vehicle. A quick-thinking crew member is, nearly always, able to contain the damage, often at the cost of his own life.

Stories about commercial, electricity-generating nuclear power plants are also generally optimistic. A few even tend toward the messianic. Heroic engineers in both Poul Anderson's novel *Orion Shall Rise* (1983)

Mutants for Nuclear Power

An example of antinuclear graffiti. Dark humor is common on both sides of the ongoing debate over the safety of nuclear power plants. Here, the issue is the long-term effect of radiation emissions on nearby residents.

and Larry Niven and Jerry Pournelle's novel *Lucifer's Hammer* (1977) use power from nuclear plants to rebuild civilization after global natural disasters. *Fallen Angels*—a 1991 collaboration between Niven, Pournelle, and Michael Flynn—is set in a near-future North America in the grip of a resurgent Ice Age. Nuclear-generated heat *could* have saved the now-buried northern states had not radical environmentalists blocked construction of the necessary plants.

Some stories about commercial nuclear power plants are less ferocious in their optimism. From Lester Del Rey's "Nerves" (short story 1942, novel 1956) to *The China Syndrome* (film 1979), they typically depict accidents but focus on the plant operators' heroic efforts to contain the danger. The efforts are nearly always successful, though they often leave one or more of the heroes dead, injured, or psychologically battered. Both kinds of stories share an underlying message: Nuclear plants are safe, but only in the hands of trained professionals willing to give their lives to protect the public from the terrible forces they control.

Optimistic and pessimistic views of nuclear power plants are visible in much purer form in work designed to persuade more than entertain. Antinuclear graffiti plays on fears that radiation-leaking power plants will produce hideous mutations. One widely circulated design shows a cartoon of a grotesquely mutated family group, labeled "The Nuclear Family." Nuclear power advocates counter with their own slogans, often taking satirical aim at prominent antinuclear figures. "Nuclear plants," proclaims one, "are built better than Jane Fonda." Another compares the death tolls of Senator Edward Kennedy's career-staining 1969 car accident and the 1979 near disaster at a Harrisburg, Pennsylvania, nuclear plant: "Chappaquiddick—1, Three Mile Island—0; Go Nuclear."

Decisions about the use of nuclear energy are shaped as much by political and social concerns as by scientific ones. The nonscientific issues are, however, rooted in the scientific knowledge summarized in Einstein's equation $E = mc^2$: nuclear energy is the most powerful force that can be brought under human control. To some, efforts to control and use it represent limitless opportunity. To others, they represent unconscionable risk.

Related Entries: Mutations; Radiation

FURTHER READING

Badash, Lawrence J. *Scientists and the Development of Nuclear Weapons*. Humanities Press, 1995. Brief introduction, for nonspecialists, to events up to 1963.

Boyer, Paul L. *By the Bomb's Early Light: American Thought and Culture at the Dawn of the Atomic Age*. Pantheon, 1985. Definitive study of American reactions to nuclear weapons from 1945 into the 1950s.

Del Sesto, S.L. "Wasn't the Future of Nuclear Energy Wonderful?" Chap. 3 of Joseph J. Corn, ed., *Imagining Tomorrow* (MIT Press, 1986). Examines early, overoptimistic 1940s and '50s forecasts of the commercial potential of nuclear power.

Kaku, Michio, and Jennifer Trainer. *Nuclear Power: Both Sides*. Norton, 1983. An evenhanded, journalistic account of the controversy over nuclear power plants, outlining the positions of both sides.

Winkler, Allan M. *Life under a Cloud: American Anxiety about the Atom*. University of Illinois Press, 1999. Broad, concise survey of the American public's reactions to both nuclear weapons and nuclear power since 1945. Less depth than Boyer but more chronological range.

Chimpanzees

Chimpanzees, like gorillas and orangutans, are members of a family of primates known as the great apes—humankind's closest living relatives. They are forest dwellers who live primarily on fruit but also eat nuts, insects, the meat of small animals, and the young of larger ones. Chimps are known for their high intelligence and complex social structure. They hunt cooperatively, share food, and respond in coordinated ways to approaching enemies. They communicate vocally in the wild, and individuals have been taught in captivity to communicate with humans through gestures and signs. Chimps make and use tools as humans do and, like humans, pass on the knowledge of *how* to make tools to their young. They also, according to recent studies, commit premeditated acts of violence against one another. This violence includes both the murder of individual chimps by rivals and organized warfare between communities competing for foraging territory. The killings sometimes, but not always, end in cannibalism.

For members of a species so closely related to humans, so complex in its social organization, and so clearly intelligent, chimpanzees get little respect in popular culture. Whales and dolphins are admired for their grace, dogs for their loyalty, and horses for the working partnerships they form with humans. Chimps, however, are simply "cute." Popular culture has, for a century, consistently portrayed chimps as inconsequential: as clowns, sidekicks, and childlike companions. It has only rarely allowed them the kind of active, independent roles routinely assigned to dolphins (*Flipper*), dogs (*Lassie*), or horses (*The Black Stallion*).

Chimps' prominent facial features, long limbs, and quick movements make them inherently amusing to most human audiences, especially if the chimps mimic human dress and activities. Chimp entertainment acts based on this principle, staples of circuses and stage shows in the first half of the twentieth century, easily made the transition to television. On television, a new variety of chimp act developed: weekly series whose casts consisted entirely of chimps wearing human clothing and "speaking" lines dubbed by human actors. All-ape series from *Lancelot Link: Secret Chimp* in the early 1970s to *The Chimp Channel* in the late 1990s relied on low-comedy standards like pratfalls and bodily-function jokes to sustain the story and depended on the novelty of performing chimps to make the timeworn gags funny. Their chimp stars were, therefore, as interchangeable as the chimps in a circus or stage act. They were funny as members of a species, not as individuals.

Most chimps featured in film and on television play specific characters and have specific roles in the on-screen story, but the characters are stereotyped and the roles limited. Most are comic sidekicks: to host Dave Garroway in the early years of the *Today Show* (1954–1957), to a veterinarian in *Daktari* (1966–1969), to a trucker in *B.J. and the Bear* (1979–1981), and to a globe-trotting family in *The Wild Thornberrys* (1999–   ). Others are surrogate children: to a befuddled psychologist in the film *Bedtime for Bonzo* (1951), for example, or to an otherwise ordinary suburban couple in TV's *The Hathaways* (1961–1962). When confined to such roles, chimps have little to do with the main story being told. They may be whirlwinds of activity on screen, but as is true of human sidekicks and human children, their actions are digressions from the plot rather than steps toward its resolution. *Lassie Come Home* (1943) is about a dog and *King Kong* (1933) about an ape, but *Bedtime for Bonzo* is about Ronald Reagan's character and his comic attempts to cope with a mischievous chimp.

Exceptions to this pattern—tales where chimpanzee characters take center stage and shape their own destinies—are rare but significant. One of the first, Robert A. Heinlein's story "Jerry Was a Man" (1947), is the story of a genetically enhanced chimp who sues his human employer for pay and benefits matching those earned by human workers. *Project X* (1987) focuses on a young air force pilot who tries to save chimps slated for a lethal experiment, but it makes the chimps into active coconspirators. *Conquest of the Planet of the Apes* (1972), fourth of five films in the series, chronicles a near-future (1991) revolt by enslaved apes against their human masters. The leader of the revolt, Caesar, is a highly evolved chimpanzee whose parents traveled back in time from the ape-dominated Earth of the 3900s A.D. He is, thanks to a complex time loop, the architect of the world into which his parents would be born. The fierce, efficient mass violence of the revolt (modeled on the Watts riots of 1965) seems far more plausible as chimp behavior now than it did in 1971. Caesar may yet prove to be the truest fictional representative of his species.

Related Entries: Gorillas; Intelligence, Animal; Time Travel

FURTHER READING

Fouts, Roger, et al. *Next of Kin: My Conversations with Chimpanzees*. Bard Books, 1998. Survey, for nonspecialists, of research on chimpanzee intelligence and communications.

Goodall, Jane. *Through a Window*. Houghton Mifflin, 1990. Houghton Mifflin, 2000. The foremost observer of chimpanzee behavior in the wild summarizes thirty years of fieldwork.

Landau, Virginia, et al., eds. *Chimpanzoo*. March 2000. The Jane Goodall Institute. 5 May 2000. <http://chimpanzoo.arizona.edu/>. Information on chimpanzees generally and research on chimpanzee behavior (including communication) in captivity.

Primate Information Network. "Common Chimpanzee (*Pan troglodytes*)" 8 April 2000. Wisconsin Regional Primate Research Center, University of Wisconsin–Madison. 7 December 2001. <http://www.primate.wisc.edu/pin/factsheets/pan_troglodytes.html>. Ten authoritative fact pages on chimpanzee anatomy, behavior, ecology, and communication, plus a bibliography.

Cloning

A clone is a genetically identical copy of an organism, created when the nucleus of a single cell in the parent's body divides itself. Single-celled organisms such as bacteria, which reproduce by dividing, clone themselves naturally. Natural cloning is rare among multicelled animals, since it limits genetic diversity and tends to hasten extinction. Identical twins, formed when a fertilized egg divides in the womb, are also clones—genetically identical but produced by sexual rather than asexual reproduction and carrying the genes of two parents rather than one.

Cloning in the laboratory involves taking a cell from the organism to be cloned, removing its nucleus, and transferring the DNA to an egg cell from which the DNA has been removed. The egg is then implanted in the womb of a surrogate mother and brought to term. Laboratory cloning became possible in the 1980s, but only by using cells taken from an embryo—cells that had not yet differentiated and specialized to form particular organs. The crucial breakthrough came in the summer of 1996, when a team of scientists working for Scotland's Roslin Institute produced a healthy clone using a cell taken from the udder of an adult sheep. The clone, Dolly, became a worldwide celebrity and a catalyst for intense debates over ethical and public policy issues. Successful cloning of cattle, pigs, and mice followed, and on 25 November 2001 scientists at Advanced Cell Technologies in Cambridge, Massachusetts, announced the successful cloning of human embryos.

Dolly the sheep was the Roslin Institute's first success in 267 attempts, and Advanced Cell Technologies' cloned embryos did not grow past the eight-cell stage. Those successes, however, created the widespread conviction that the ability to clone at will is only a matter of time. The United States and Britain placed moratoria on human cloning research within a year of the birth of Dolly. Australia and many European countries instituted various regulations of their own. Japan passed legislation in December 2000 making human cloning a crime, and the United States is now considering a similar ban. Popular culture reflects the ethical, religious, and legal concerns that drove this legislation. Long before Dolly, it established extravagant expectations of what cloning could do. The remaining sections of this essay discuss popular culture's three most common images of cloning.

CLONING AS DUPLICATION

Depictions of cloning in popular culture often make two invalid assumptions. The first is that a clone is not just a genetic duplicate of its parent but an identical physical duplicate as well. The second is that a human clone will be, when it emerges from the laboratory, the same biological "age" that its parent was at the time of cloning. Popular culture tends, therefore, to imply that cloning inevitably produces exact duplicates.

The flaw in the "same age" assumption is obvious. Human clones would, like Dolly and other animal clones, grow from embryos—albeit genetically altered ones. A man who was X years old when his cloned offspring was born would, like any other parent, always be X years older than his child. The flaw in the "physical duplicate" assumption is more subtle. Genes shape an individual's physical appearance, but so do environmental factors, such as nutrition and disease. A clone, like a child conceived sexually, would grow in a different womb and grow up in a different physical environment. The clone would strongly resemble, but not precisely duplicate, its single parent,

Strict attention to biological reality would, of course, preclude many of the dramatic possibilities that make clones interesting subjects for stories. Fascination with twins and duplicates runs deep in Western culture, from Shakespeare's *Comedy of Errors* and Dickens's *Tale of Two Cities* through *The Prince and the Pauper* and *The Parent Trap*. Photocopy-style cloning adds the possibility of creating duplicates at any time—and in any quantity—the duplicator desires. Popular culture tends, not surprisingly, to choose drama over accuracy.

Ben Bova's science fiction novel *The Multiple Man* (1976) begins with the discovery of a dead man physically identical to the still-living president of the United States. It builds to the revelation that the president has been cloned, in order simultaneously to appear in public and work behind closed doors. The 1996 film *Multiplicity* uses the same concept as the basis for farce. Doug Kinney (Michael Keaton) has himself secretly cloned so that he can literally be in two places at once. The deception works, fooling his friends and even his wife, because the cloned Dougs are physically identical to the original. *Quark*, a short-lived 1978 TV comedy series about the crew of a interstellar garbage scow (barge), features a pair of physically identical (and identically dressed) characters named

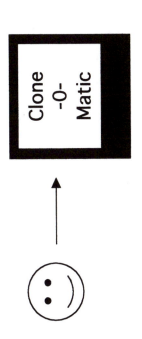

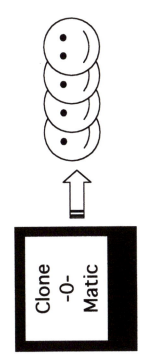

The popular image of human cloning. Popular culture often misrepresents cloning as the biological equivalent of photocopying: a means of producing, instantaneously, copies identical in every detail to the original.

Betty 1 and Betty 2. One Betty is a clone of the other, but each, in a running joke, firmly insists that she is the original.

The two Bettys are a relatively rare example of individuals and their clones being given identical personalities and mannerisms. The drama in *The Multiple Man* and the biggest laughs in *Multiplicity* depend on the fact that their clones have individual personalities. The ex-Nazis who attempt to clone Adolf Hitler in Ira Levin's novel *The Boys from Brazil* (1976) fail because they cannot recreate the social environment that shaped him. The young hero of Mike Resnick's *The Widowmaker* (1996), cloned from the dying body of the galaxy's most-feared bounty hunter, inherits keen eyesight and quick reflexes from his "father" but not the cynical detachment born of decades in the business. Stories about clones are, in fact, more likely than much of popular culture to appreciate the role of social forces in shaping individual personalities.

CLONING AND MASS-PRODUCED HUMAN BEINGS

Human cloning, according to many post-Dolly commentators, is frightening because of its potential for dehumanization. The mass production by cloning of hundreds of identical human beings is a particularly potent symbol of this kind of dehumanization. It makes literal what has for most of the industrial age remained metaphorical: the reduction of human beings to "product"—to interchangeable spare parts. Popular culture's depictions of mass-production cloning have, not surprisingly, been strongly and uniformly negative.

The mass cloning attempted by Dr. Josef Mengele in *The Boys from Brazil* requires a small army of women to carry the small army of Hitler clones to term. The women of *The Boys from Brazil* are reduced to machines, living incubators for the clones. The clones themselves are, in a different sense, equally dehumanized. They are created and valued not for their individuality but for their hoped-for resemblance to the long-dead *führer* whom Mengele and his henchmen wish to restore to the world. The TV series *Space: Above and Beyond* (1995–1996) includes, as part of its social background, a class of clones called "in vitros," or "tanks," (for the giant laboratory vessels in which they are grown). The "tanks" were designed as cannon fodder: disposable soldiers for a brutal war in which they, displaying unexpected individuality, refuse to fight. *The Resurrection of Zachary Wheeler*, a 1971 film, takes the equation of cloning and dehumanization to its logical conclusion. In it, a secret clinic rebuilds the bodies of horribly injured patients by cloning them and using the clones for spare parts.

CLONING THE DEAD

Cloning, as depicted in popular culture, is full of scientists who "defy the natural order of things" by attempting to create living, breathing copies of the dead. Popular culture, from W.W. Jacobs's "The Monkey's Paw" (1902) to Stephen King's *Pet Sematary* (1983) has long viewed attempts to raise the dead as ill advised. Combined with cloning, it is certain proof that a scientist possesses an unhealthy ego, perhaps even sees himself as God.

Ian Malcolm, the mathematician who serves as the voice of restraint in *Jurassic Park* (novel 1990, film 1992) chides entrepreneur John Hammond for cloning dinosaurs simply to turn a profit. It would be different, Malcolm argues in the film, if Hammond were resurrecting the condor: they were brought to extinction by human action and so, implicitly, resurrecting them by human action would provide symmetry and balance. Dinosaurs, Malcolm insists, are another matter: nature "meant" for them to die, and it is the height of arrogance for humans to attempt to overturn that judgement. Harry Wolper, the eccentrically brilliant scientist who sets out to clone his dead wife in *Creator* (novel 1980, film 1985) cheerfully admits that he is playing God. Indeed, he relishes the role. One day, he proclaims to his astonished graduate student, "I'm going to look into my microscope and find God looking back, and the one of us that blinks first is going to lose! . . ."

Nature ultimately bests both Wolper and Hammond. Wolper is unable to clone his wife but develops a "natural" attraction to another woman, who fills the void left by her death. Hammond's dinosaur theme park collapses into chaos, and he either learns his lesson (in the film) or is eaten by his creations (in the book). The clone makers of *The Boys from Brazil, The Widowmaker,* and *Space: Above and Beyond* also fail to achieve their goals: they can dictate the physical form of their clones but not their behavior. Even the well-intentioned cloning depicted in *Multiplicity* and *The Multiple Man* ultimately fails: it is designed to deceive, but the limits of cloning reveal the deception.

The fact that fictional attempts at cloning nearly always go awry reflects one side of popular culture's divided attitude toward human attempts to control nature. Manipulating the nonliving world—damming a river, diverting a lava flow—is heroic. Manipulating the living world—cloning humans, growing giant animals—is evidence of a dangerously unstable mind.

Related Entries: Genes; Genetic Engineering; Organ Transplants; Superhumans

FURTHER READING

"Breaking News: Meet Dolly . . ." Biospace.com. 9 December 2000. <http://www.biospace.com/b2/whats_new/dolly.cfm>. Includes chronologically organized links to news stories, a bibliography, and excerpts from published commentaries.

"Cloning Special Report: Cloning and Stem Cells" *New Scientist*. 13 March 2002.

28 March 2002. <http://www.newscientist.com/hottopics/cloning/>. An indispensable guide to cloning: news stories, frequently asked questions, scientific background, and high-quality links.

Kolata, Gina. *Clone: The Road to Dolly and the Path Ahead.* Morrow, 1999. Places the cloning of Dolly (and its likely impact) in the context of earlier attempts, alleged attempts, and hoaxes.

Pence, Gregory E., ed. *Flesh of My Flesh: The Ethics of Cloning Humans.* Rowan and Littlefield, 1998. Post-Dolly commentaries by leading scientists.

Wilmut, Ian, Keith Campbell, and Colin Tudge. *The Second Creation: Dolly and the Age of Biological Control.* Harvard University Press, 2000. The biology, mechanics, and possible future of cloning, by those involved.

Comets

Comets are small celestial bodies made of dust and ice—"dirty snowballs," in the words of astronomer Fred Hoyle. They come from reservoirs on the outermost fringes of the solar system: the Kuiper Belt, several times more distant than Pluto; and the Oort Cloud, 1,000 times more distant than Pluto. The comets we see from Earth have been knocked loose from these reservoirs and pulled into long, narrow elliptical orbits around the Sun. Small amounts of ice vaporize as the comet approaches the sun. The vapor forms a "coma" of gas and dust around the comet's still-solid nucleus and a "tail" that streams out behind it. Comets' tails can be spectacular—the tail of Halley's Comet covered a ninety-degree arc of the night sky in 1910—but they are very tenuous. A cubic kilometer of tail contains less matter than a cubic millimeter of air.

Every comet leaves a trail of dust and ice particles behind it as it passes through the solar system; occasionally, a comet disintegrates entirely. A comet (or comet fragment) may have caused the massive explosion at Tunguska, Siberia, in June 1908. Comet impacts may also have been responsible for some of the periodic mass extinctions that punctuate the history of life on Earth.

Comets have often been regarded as omens—signs of triumph or disaster to come. *What* they signify is not always clear. A comet hung in the skies over Western Europe in 1066 as William of Normandy prepared to fight Harold Godwinson for the right to rule England. William must have seen it, in retrospect at least, as a good omen: he won the battle, the throne, and the name "William the Conqueror." Harold, dying on the battlefield with a Norman arrow through his right eye, doubtless had his own opinion.

Halley's Comet, which has returned every seventy-six years for the last 2,000, has acquired a reputation as a signpost in human lives. Even in the nineteenth century, seventy-six years was a long—but not

Halley's Comet in 1910, drawn by Elizabeth Shippen Green for the 21 May 1910 issue of *Harper's Weekly*. Many observers, primed by pictures like this one, were disappointed by the comet's lackluster appearance in 1986. Courtesy of the Library of Congress.

improbably—long life. Most people would see the comet once in their lifetime, but someone born in a year when it appeared might see it twice. Mark Twain, noting that he'd "come in with the comet" in 1835, speculated that he'd "go out with the comet" when it returned. He did just that, dying in 1910 as the comet made an unusually spectacular appearance. Halley's Comet also provided bookends for the life of Mississippi-born writer Eudora Welty, who saw it as an infant in 1910 and as an old woman in 1986. Mary-Chapin Carpenter's song "Halley Came to Jackson," from the album *Shooting Straight in the Dark* (1990), was inspired by the story. It uses the comet not as an omen of impending death but as a sign that Welty had lived a long, full life. Thirty-seven members of the Heaven's Gate cult committed suicide in March 1997, believing that the appearance of comet Hale-Bopp was a signal for them to "shed their bodies" and rendezvous with an alien spacecraft they believed to be following the comet.

Halley's aside, comets appear most often in recent popular culture as agents of death and destruction on a planetwide scale. Science-fiction writers Larry Niven and Jerry Pournelle used a comet to wipe out much of the Earth's population in their novel *Lucifer's Hammer* (1979). Its impact in the Pacific Ocean produces a massive tsunami and one of the book's most memorable images: a dedicated surfer getting the last and greatest ride of his career on the face of the world's biggest wave. The comet in the 1984 film *Night of the Comet* causes more varied, if less plausible, mayhem: most of those not killed outright by its explosion are turned to zombies. *Deep Impact*, one of two extraterrestrial impact movies from the summer of 1998 (for the other, *Armageddon*, see "Meteorites"), has the most menacing fictional comet of all. If it strikes the earth intact, it will trigger a new mass extinction. If only a piece of it strikes, hundreds of thousands will die, but civilization will survive.

The fictional comets, for all their modern trappings, do what comets have always done in popular culture: they signal imminent, world-altering events. They are, as usual, maddeningly vague omens. The fictional characters, as did Kings William and Harold in 1066, know that changes are coming but not what the changes will be.

Related Entries: Extinction; Meteorites

FURTHER READING

Burnham, Robert. *Great Comets*. Cambridge University Press, 1998. Focuses on large, spectacular comets and their cultural impact.

Gropman, Donald. *Comet Fever: A Popular History of Halley's Comet*. Simon and Schuster, 1985. Focuses on the impact of Halley's spectacular 1910 appearance.

Plait, Phil. "Phil Plait's Bad Astronomy: The Home Page." 14 August 2001. 15 August 2001. <http://www.badastronomy.com/>. Outlines, in detail, astronomical errors in movies and TV shows.

Computers

Computers are machines for electronically storing and manipulating data. Any computer, no matter how large or powerful, is a system of interconnected components: a central processing unit (CPU) linked to input, output, and storage devices. The physical components are known as "hardware," the sets of instructions that control them as "software." Computers come, for the purposes of this entry, in two basic types: large "mainframes" shared by multiple users who access them through terminals, and desk or laptop "personal computers" (PCs) designed to serve one user at a time. PCs may also be linked to other PCs, to a nearby mainframe, or—by modems and phone lines or more sophisticated means—to a distant mainframe that provides a gateway to the global Internet.

COMPUTERS AND PEOPLE

Computers' value lies in their ability to work faster and more accurately than humans can, without direct human supervision. Once furnished with a set of instructions and a set of data to which to apply them, a computer becomes one of the few human-made tools that can act entirely independently of its user. They are independent, however, only within narrow limits. Human users must supply data in a form that they can recognize, as well as sets of instructions that are internally coherent and capable of producing the desired results. Even trivial errors in the data or the instructions applied to it can render the computer useless. The slogan "garbage in, garbage out," coined early in the computer age, reflects a fundamental truth: computers are only as good as the fallible humans behind them.

ENIAC, widely regarded as the first true electronic computer, made its debut at the University of Pennsylvania in 1946. The next fifty-five years of computer history embodied two trends: increasing efficiency and growing democratization. Computers became steadily more powerful, compact, and reliable as vacuum tubes gave way to transistors in the 1950s and integrated circuits in the 1960s. Advances in integrated circuits and the microprocessors they make up have continued the more-power-in-less-space trend since then. Hundred-dollar programmable calculators, according to one widely quoted anecdote, are now more powerful than the onboard computers that guided Apollo spacecraft to the Moon a generation ago. The same trend made it possible by the mid-1970s to put a powerful computer in a typewriter-sized box. The advent of such machines began the democratization of computers. The development of intuitive controls—notably the Macintosh and Windows operating systems in the mid-1980s—extended the process. The explosive growth of the Internet in the 1990s completed it, drawing millions of new users into their first extended interactions with computers and creating new markets for low-cost, easy-to-use machines.

Computers affected everyday life even when they were big, expensive, and rare. Their transformation into everyday tools, as familiar as telephones or television sets, has made their effects more visible and more pervasive. Portrayals of computers in popular culture reflect the depth and complexity of human users' half-century relationship with them. The portrayals play, often simultaneously, on the extravagant hopes and the intense anxieties that computers generate.

FICTIONAL MAINFRAMES

"Computer" and "mainframe" were synonymous in popular culture from the 1950s through the mid-1980s. Big, powerful machines became symbols of the big, powerful organizations that bought and used them. Replacement by a machine, a fate once feared only by blue-collar workers, became a white-collar nightmare as well. The threat, though powerful, remained unspecific. It belonged to the same undefined near future as the "paperless office," and the best-known examples of office workers replaced by computers remained fictitious or apocryphal. "The Freshmen down at Yale," a student song from the early 1960s about enforced chastity at the then-all-male university, explained the sex-free life of the university bursar this way: "It's not that he's so clean/He's an IBM machine."

Popular culture also gleefully cataloged mainframes' faults: the inexplicable breakdowns they suffered, the constant monitoring they demanded, and the absurd results they produced when given flawed data or badly written programs. "Do Not Fold, Spindle, or Mutilate," the stern warning printed on the punched cards used to input data, became an irony-laden catchphrase—a commentary on organizations more solicitous of their computers than of their human employees. The 1962 film

That Touch of Mink, a romantic comedy set in a corporate office building, features a computer that goes berserk and spews its carefully sorted punch-cards across the room. A mid-1970s feature in *Mad* magazine urged frustrated businessmen to "take an axe" to error-prone computers that scrambled payrolls and garbled orders.

EMIRAC, the immense "electronic brain" featured in the 1957 film comedy *The Desk Set*, has both the exaggerated capabilities and the exaggerated flaws that popular culture routinely attributed to mainframes. EMIRAC can understand queries typed in plain English and ask—again in plain English—for clarification. It can also answer questions ("How much damage is done to forests annually by the spruce bud worm?") that require it to independently select relevant data and analyze their significance. These features place EMIRAC well ahead of the most sophisticated search engines in general use half a century later, but they mask serious flaws. Asked for information on a subject, EMIRAC apparently prints the first approximately relevant piece of data it finds in its databanks. A mistyped request for information on Corfu brings forth page after page of an English ballad whose title contains the word "curfew."

Mainframes' steadily expanding role in processing routine government and corporate data led by the early 1960s to cautionary tales of societies that allowed their computer servants to become their masters. They appeared on episodes of TV series such as *The Twilight Zone* ("The Old Man in the Cave," 1963) and *Star Trek* ("Return of the Archons," 1967) and in movies such as *Logan's Run* (1976). A variation on the story—*2001: A Space Odyssey* (1968) is a classic example—places human characters in a confined space controlled by a malevolent mainframe. A third version involves humans who—again, too trusting of computers—place nuclear weapons under their direct control. Catastrophe follows mechanical failure in *Fail-Safe* (novel 1962, film 1964). In the machine's conscious action—as in *Colossus: The Forbin Project* (novel 1966, film 1970) and the *Terminator* movies (1984, 1991)—the machines' conscious actions are responsible.

The message of these stories is the office worker's fear of replacement-by-computer on a grand scale. Both rest on the pervasive assumption that mainframes are (or will soon be) so "intelligent" that they *could* replace humans. A computer capable of doing so—Colossus and *2001's* HAL 9000, for example—would have to mimic humans' flexible, adaptable intelligence. It would have to be able to think beyond its programming, adapting its thought patterns to new situations and new forms of data. It would, in other words, have to clear a very tall barrier, the base of which real-world computer designers have only begun to probe.

FICTIONAL PERSONAL COMPUTERS

Fictional PCs, unlike fictional mainframes, closely mimic the functions and abilities of their real-world counterparts. The major differences—greater speeds, fewer crashes, and perfect interfacing—are concessions to storytelling efficiency. The lead characters in the 1998 film *You've Got Mail* send and receive e-mail without busy signals, overburdened servers, or dropped connections, because such real-world intrusions would complicate their love story in dramatically uninteresting ways.

Equally important, fictional PCs are not independent entities but tools that remain under their users' control at all times. Far from threatening to reduce their users to numbers in a database, as fictional mainframes do, they allow users to assert and protect their individuality. Fictional PCs are, in their most visible roles, tools of people who stand outside "the system" and may even be trying to subvert it. The title character in Peter Zale's comic strip *Helen, Sweetheart of the Internet* (1996–   ) uses her PC and genius-level intelligence to take elegant revenge on her corporate enemies. The secret-agent heroes of the film *Mission: Impossible* (1996) use a laptop to break into the CIA's mainframe in order to gather evidence against a traitor. A computer genius saves the Earth from implacable aliens in the film *Independence Day* (1996) by the wildly improbably method of using a laptop to infect the alien invasion fleet's mainframes with a super-destructive computer virus. "Cyberpunk," a subgenre of science fiction that emerged in the mid-1980s, created a now-mainstream vision of the future as a dark, chaotic world in which morally ambiguous heroes live by their wits and their cyber-skills. Skill with a PC became a virtual job requirement for edgy rebel-heroes of the 1990s, from the storm chasers of Bruce Sterling's novel *Heavy Weather* (1994) to the guerillas of *The Matrix* (1999).

Apple Computer, Inc., plays brilliantly with these images in its advertising, positioning its products as the tools of *true* individualists. The commercial that introduced the Apple Macintosh in the winter of 1983–1984 shows a vibrant young woman's spectacular act of rebellion against a totalitarian dictator. It ends with the reassurance that because of Macintosh, "1984 won't be like *1984*." A later, print-based campaign (1997–   ) associated the Apple logo and the slogan "Think Different" with images of Einstein, Gandhi, Amelia Earhart, and other noted innovators. Apple's current TV spokesman, actor Jeff Goldblum, trades on his history of playing edgy, unconventional characters—including the laptop-wielding savior of humankind in *Independence Day*.

Related Entries: Androids; Intelligence, Artificial; Robots

FURTHER READING

Ceruzzi, Paul E. *A History of Modern Computing*. MIT Press, 2000. Surveys the history of computer hardware from ENIAC (1946) to the rise of the World Wide Web in the mid-1990s.

Comer, Douglas E. *The Internet Book: Everything You Need to Know about Computer Networking and How the Internet Works*. 3rd ed. Prentice-Hall, 2000. A nontechnical, jargon-free survey of the history and the current state of the Internet, focusing on the technology involved.

Levy, Stephen. *Hackers: Heroes of the Computer Revolution*. 1985. Penguin, 2001. Classic, novelistic history of the origins of the personal computer revolution (late 1950s–1984) and the "hacker philosophy" of cooperation, shared knowledge, and individual autonomy.

Stoll, Clifford. *High Tech Heretic: Reflections of a Computer Contrarian*. Anchor Books, 2000. Twenty-three critical essays on the social and cultural impact of computers and the Internet, written by a computer engineer determined to challenge conventional wisdom.

White, Ron. *How Computers Work: Millennium Edition*. Que, 1999. A nontechnical, graphics-laden introduction to the hardware that makes up personal computers and peripheral devices.

Cryonics

"Cryonics" is an experimental medical procedure in which the bodies of the recently dead are stored at very low temperatures for later revival. It is not related, except in its focus on cold, to "cryogenics"—a branch of physics and engineering focused on the production and study of low temperatures. Roughly 100 people have been frozen since the late 1960s, and roughly 1,000 more have arranged to be frozen once they are declared legally dead. The justification for cryonic storage rests on the principle that today's lethal diseases will be curable in the future. Preserved, their medical histories carefully documented, the frozen dead can be revived and restored to health once medical science has advanced enough to reverse the damage that killed them and the additional damage done by freezing. Cryonics advocates acknowledge that successful revivals are unlikely in the foreseeable future but note that individuals in cryonic storage, being legally dead already, have nothing to lose by waiting.

Popular culture seldom treats cryonics in the context of the present. When it does, cryonics is seldom the point of the story. Thawing out an alien monster from the polar ice in John W. Campbell's "Who Goes There?" (1938) and its movie adaptation *The Thing* (1951, 1982) is a prelude to a suspense-horror story. Thawing out a 40,000-year-old Neanderthal man from a glacier in the movie *Iceman* (1984) is a prelude to explorations of what makes us human. Movies like *Sleeper* (1973) and *Forever Young* (1992) use cryonics to establish Rip Van Winkle–style plots—comic in one case, romantic in the other. They treat the dilemma of a frozen-and-thawed hero who wakes up to find the world made alien by the passage of time.

Stories about cryonic preservation that are set wholly in the future also take the process for granted, but only to a point. They postulate that the basic process is sound and that it has become standard under certain narrowly defined circumstances: for people with currently incurable diseases, for example, or for space travelers preparing for long voyages.

They also tend, however, to assume that cryonics has unforeseen risks and drawbacks that complicate the lives of those involved in it. Those complications are, frequently, central to the story.

Often, the complications are medical: lingering effects of freezing that were unforeseen or assumed (wrongly) to be of little consequence. A doctor and a patient fall in love, in Spider Robinson's short story "Antinomy" (1978), a few months before the latter enters cryonic storage. The fact that the process will erase six months of her memories, including their romance, is a minor problem . . . until an unexpected breakthrough allows her to be revived shortly after she is frozen. The interstellar travelers of the novels *The Legacy of Heorot* (1987) and *Beowulf's Children* (1995) awaken from cryonic storage with their bodies intact but their brain functions impaired. They can function adequately on a day-to-day basis but can no longer trust that their judgement is entirely sound. They conclude that group decision making will prevent any one person's impaired judgement from endangering the colony, and they design their new society accordingly.

Equally often, however, the complications are external: changes in society that imperil characters in cryonic storage who would otherwise have nothing to fear from the process. Three scientists frozen for a long voyage to Jupiter in the film *2001: A Space Odyssey* (1968) die when a conscious member of the crew (HAL 9000, the ship's computer) goes berserk and shuts down their life-support systems. A terminally ill patient frozen in order to wait for a cure awakens in Mike Resnick's novel *The Widowmaker* (1992) to the unwelcome news that the money set aside to pay for his maintenance costs is nearly exhausted. He faces a terrible choice: stay thawed and die of his still-incurable illness or raise the huge sum necessary to pay for additional freezing and storage. Cryonic storage is so common in the world of Larry Niven's short story "The Defenseless Dead" (1973) that the frozen dead (called "corpsicles") outnumber the living. The dead are placed in peril when the living, who have all the political power, begin to view them not as people but as a resource.

Major scientific and technological breakthroughs invariably have unintended consequences. Modern appliances, for example, gave the average American household fresher food and cleaner clothes but—contrary to everyone's expectations—led to women's spending more time on housework rather than less. Popular culture has often been a key venue for exploring the unintended consequences of new discoveries, and its treatment of cryonics shows it operating in that crucial role.

Related Entries: Longevity; Space Travel, Interstellar

FURTHER READING AND SOURCES CONSULTED

Alcor Life Extension Foundation. "What We Do: A Short Introduction to Cryonics." 1998. 15 December 2000. <http://www.alcor.org/01b>. A short, nontechnical introduction to cryonics and the logic behind it, from a leading advocacy group.

Bridge, Steve, ed. "Cryonics." The Open Directory Project. 15 December 2000. <http://dmoz.org/Science/Biology/Cryobiology/Cryonics/>. A thorough, categorized, briefly annotated "bibliography" of Web sites dealing with cryonics.

Cowan, Ruth Schwartz. *More Work for Mother: The Ironies of Household Technology from the Open Hearth to the Microwave.* Basic Books, 1983. A landmark study of the "Law of Unintended Consequences" at work in the history of technology.

Cyborgs

Cyborgs are composite beings: part biological, part mechanical. They retain parts of their human bodies and all of their human consciousness. Unlike androids, which are machines that have human form, cyborgs are humans whose original, biological components have been selectively replaced with mechanical ones.

Body-part replacement is, in a limited sense, an old and well-established process. False teeth and prosthetic limbs are centuries old, and replacements for degraded joints with teflon and steel is now an established medical practice. Mechanical devices designed to augment body parts are also common, ranging from eyeglasses to implanted pacemakers. The word "cyborg," a contraction of "cybernetic organism," was coined in the early 1960s to describe a more complex blend of the biological and mechanical. Cyborgs, according to the most common definition, have mechanical parts controlled by computers that interface with the brain and nervous system. Advanced prosthetic limbs now in development represent the first steps toward this goal. When perfected, they will respond to a user's nerve impulses as smoothly as the flesh-and-blood limbs they replace.

Most people, in the unlikely event that they were asked to describe a cyborg, would probably imagine a person whose mechanical parts equaled or even outnumbered their biological ones. Beings meeting that description exist only in fiction, but there they are common.

Fictional cyborgs who act primarily as action-adventure heroes gain superhuman powers from their mechanical parts at little personal cost. Steve Austin, rebuilt with "bionic" parts after a catastrophic plane crash, has good reason to agree with the voice-over that began each episode of the 1974–1978 television series *The Six Million Dollar Man*: "We can make him better than he was: better, stronger, faster." Thanks to his bionic legs and right arm, Austin can run as fast as a car, jump dozens of feet straight up, and lift hundreds of pounds without breaking a sweat. His computer-enhanced left eye works as a variable-magnification telescope and range finder—a small, portable spy satellite. His mechanical parts rarely malfunction unless the plot demands it, and after the pilot episode, the psychological scars of his near-fatal accident and high-tech resurrection quickly fade. If being part machine bothers him, he seldom shows it.

Not all fictional cyborgs enjoy such uncomplicated lives. Alex Murphy, the hero of *RoboCop* (1987) and two sequels, and Darth Vader, the arch-villain of the *Star Wars* saga, gain physical strength by becoming cyborgs but also lose some of their humanity. The metal masks that hide their faces and the distorted, mechanical sounds of their voices suggest that they are in danger of *becoming* machines. Their cold, emotionless behavior hints at the same danger—as if their mechanical parts were demons taking possession of their soul. The Borg, a cyborg race that periodically rampages through the *Star Trek* universe, are the ultimate practitioners of this kind of possession. They reproduce, vampirelike, by transforming their victims. Humans captured and "assimilated" by the Borg are violated twice: first physically, by the insertion of Borg-designed mechanical implants into their bodies, and then psychologically, by the resulting submergence of their individualities in the Borg's group mind.

Stories about cyborgs losing their humanity play on deep-seated fears but nearly always have happy endings. Murphy's transformation into Robocop veils but does not erase his human consciousness, and in time his friend Anne Lewis is able to penetrate the veil and reawaken his humanity. Vader is redeemed when, at the climax of *Return of the Jedi* (1983), his deeply buried love for his son Luke Skywalker resurfaces and turns him against the evil emperor. Even the attacks of Borg are, in time, shown to be reversible. Capt. Jean-Luc Picard, captured and assimilated by them in *Star Trek: The Next Generation*, is freed from their technological bondage by surgery and from their psychological bondage by the love of his family and crew. Seven-of-Nine, a woman in her early thirties who had been assimilated by the Borg as a child, is gradually nursed back to humanity by Capt. Kathryn Janeway over several seasons of *Star Trek: Voyager*.

It is no coincidence that for all four cyborgs the key to redemption is a loving relationship with another person. Our society is deeply ambivalent about technology: delighted by the power it gives us, but concerned about its potential to separate us from each other. Cyborgs—humans tied to machines in the most intimate way imaginable—are powerful symbols of that feeling. Stories about their loss and recovery of their humanity offer a reassuring message: that technology and human feeling are not mutually exclusive.

Related Entries: Androids; Robots; Superhumans

FURTHER READING

Gray, Chris Hables, et al. *The Cyborg Handbook*. Routledge, 1996. Rigorous survey of cyborg technology already in use.

Haraway, Donna. "A Cyborg Manifesto." In *Simians, Cyborgs and Women: The Reinvention of Nature*. Routledge, 1991; also <http://www.stanford.edu/dept/HPS/Haraway/CyborgManifesto.html>. Dense, complex, classic exploration of the cultural meaning of cyborgs.

Irvine, Martin. "Cyborgs-R-Us: A Guide to the Visual History of Cyborgs." 1998. 15 August 2001. <http://www.georgetown.edu/irvinemj/technoculture/cyborgy/>. Brief but comprehensive one-page survey of cyborg images.

Darwin, Charles

Charles Robert Darwin (1809–1882) did not invent the idea that new species developed by natural processes from existing species. Rather, he reinvented it. He took what had been a fringe theory with overtones of political radicalism and transformed it into one of the central principles of modern biology, geology, and anthropology. His lifetime of research established a massive body of evidence for the reality of biological evolution. Darwin shares credit with biologist Alfred Russell Wallace (1824–1913) for discovering "natural selection," the mechanism that drives evolution (see "Evolution"). It was Darwin, however, who firmly implanted the idea of evolution in the public consciousness.

Charles Darwin was, throughout his career, deeply involved in Britain's vibrant scientific community. He rose to prominence in the Geological Society of London while still in his twenties, taking an active role in meetings and serving as secretary. Away from Britain for five years aboard HMS *Beagle*, he dispatched letters to colleagues whenever the ship made port. This stream of correspondence grew to a torrent after he returned home. Though physically isolated—living in his country house at Down, seldom venturing to London—he remained socially engaged. Specimens, observations, queries, and advice flowed steadily between Down and other centers of scientific activity. Writing his landmark 1859 work *On the Origin of Species*, Darwin drew on and synthesized the specialized expertise of hundreds of colleagues. By the time *Origin* appeared, rising young scientists like Joseph Hooker and Thomas Henry Huxley were coming to Down to consult with him.

Darwin's involvement with the world beyond Down seldom, however, extended beyond the scientific community. He did not speak publicly or write for the popular press in defense or explanation of his ideas. He never took an active role in the social and political controversies of the day, rarely even expressing an opinion on them. Most significantly, he never commented on the broader implications of his ideas: the political,

ON

THE ORIGIN OF SPECIES

BY MEANS OF NATURAL SELECTION,

OR THE

PRESERVATION OF FAVOURED RACES IN THE STRUGGLE
FOR LIFE.

By CHARLES DARWIN, M.A.,

FELLOW OF THE ROYAL, GEOLOGICAL, LINNÆAN, ETC., SOCIETIES;
AUTHOR OF 'JOURNAL OF RESEARCHES DURING H. M. S. BEAGLE'S VOYAGE
ROUND THE WORLD.'

Darwin's *On the Origin of Species* (1859). Modern readers often misunderstand the title: Darwin's theory concerned not the origin of life itself but processes by which new *forms* of life subsequently arose. Courtesy of the Library of Congress.

social, cultural, and religious dimensions that made *Origin of Species* the most controversial scientific book of the century. Others—notably Huxley and Herbert Spencer—wrote at length on such issues, but never Darwin himself.

The portrait of Darwin that emerges from popular culture transposes these two crucial aspects of his life and career. Darwin appears as a typical "lone genius" figure whose intellectual breakthroughs emerged from a one-on-one dialogue with the data. He appears, moreover, as a scientist who reveled in and enthusiastically promoted the application of his ideas to social, political, and ethical matters.

The legend of Darwin as a lone genius was largely the creation of his supporters. It now routinely appears in nonscholarly biographies and in the brief "historical" introductions and sidebars of science textbooks. Darwin's own autobiography, written for his children but widely reprinted, contributes to it. A condensed version might read as follows. The young Darwin, an indifferent student with no particular direction in life, found his true calling on the voyage of the *Beagle*. Visiting the Galapagos Islands and observing the island-to-island differences in birds and tortoises, he formulated a crucial element of his theory of evolution. Returning to England, he withdrew to his country house and worked in isolation until, in 1859, he stunned the world with *Origin of Species*. The legend reinforces an image of science popular among scientists: that of a solitary researcher whose ideas emerge from hands-on encounters with nature. It also casts Darwin in the role of a familiar folktale hero: the young man who leaves home and, after a long journey to exotic lands, returns home with a great treasure (in Darwin's case, knowledge).

The legend of Darwin as a scientist with a cultural agenda was largely the creation of his nonscientist critics. They regarded all theories of evolution, whether strictly Darwinian or not, as Darwin's intellectual progeny. All social and political movements with even a remote connection to evolution were, by extension, Darwin's handiwork. "Social Darwinism"—a late-nineteenth-century attempt to interpret social inequalities as an expression of natural law—is the most spectacular example. Darwin, who neither argued for nor believed in it, has been criticized for it by the political Left ever since. The political Right, traditional home of critics who see evolution as corrosive of faith-based morals and values, often sees Darwin as an active promoter of such corrosion. William Jennings Bryan, arguing against the teaching of evolution in U.S. public schools in the 1920s, argued that Darwinism led to a might-makes-right view of human affairs. He held Darwin all but personally responsible for German aggression in World War I. Bryan, like many present-day creationists, also painted Darwin as a religious bogeyman—one who, having lost his own religious faith, set out to rob everyone else of theirs. Jerome Lawrence and Robert E. Lee's thrice-filmed play *Inherit the Wind* (1955) uses a fictionalized 1925 court battle over the teaching of evolution to dramatize this culture clash.

Related Entries: Evolution; Religion and Science; Scientific Theories

FURTHER READING AND SOURCES CONSULTED

Appleman, Philip. *Darwin: A Norton Critical Edition.* 3rd ed. Norton, 2000. A selection of Darwin's work, with historical and scientific commentary.

Bowler, Peter. *Evolution: The History of an Idea.* Rev. ed. University of California Press, 1989. Places Darwin's work in its scientific context; principal source of the historical material in this essay.

Desmond, Adrian, and James Moore. *Darwin: The Life of a Tormented Evolutionist.* Warner Books, 1991. The most comprehensive, readable single-volume biography of Darwin.

Death Rays

Projectile weapons offer their users a crucial advantage: distance between themselves and their targets. As projectile weapons have evolved from spears and bows to guns and guided missiles, their range has steadily increased. Their striking power has also increased: a single, multiple-warhead nuclear missile can eviscerate a city. "Death rays"—a comic-book name for weapons firing energy beams—have, for nearly a century, seemed like a logical next step in the evolution of projectile weapons. Before Hiroshima, depictions of imagined "death rays" focused on their capacity for instant devastation. Later, they were re-imagined as an alternative to the indiscriminate devastation of nuclear weapons: powerful but clean and surgically precise. The laser, first developed in the 1960s, seemed to be the first real world "death ray." Characters in the TV series *Lost in Space* (1965–1968) and *Star Trek*'s first pilot episode, "The Cage" (1964), use "lasers" rather than generic "blasters" or "ray guns." Nearly forty years later, however, the capabilities of real-world lasers have yet to match those of fictional "death rays."

Lasers use a gas-filled glass tube stimulated by an electric current to produce an intense, tightly focused beam of light. They are now widely used in industry, medicine, surveying, and communications. Military applications have, until recently, been limited to guidance and range-finding systems for traditional weapons. Lasers powerful enough to damage a target directly are theoretically possible but have serious practical drawbacks. Laser beams scatter, losing cohesion and power, when they encounter the airborne particles of dust and smoke that battles often generate. Conventional lasers must be held on a target long enough to burn through it—a difficult problem, if the target can take evasive action. They also demand electricity, tethering them to power sources that are heavy, bulky, and vulnerable to damage. Laser-weapon research is now focused primarily on fixed, ground-based systems designed to destroy comparatively fragile targets such as ballistic missiles, aircraft, and (eventually) satellites. X-ray lasers, which briefly focus some of the radiation created by an exploding nuclear bomb, offer greater power but have obvious limitations.

The "death rays" of popular culture have, as their name implies, no significant civilian uses. They are purely offensive weapons, capable of neutralizing an enemy or destroying a target with a single shot. The rays have varying effects. The invading Martians of H.G. Wells's *War of the Worlds* (1901) set their targets afire with heat rays. Their successors in the film *Independence Day* (1996) induce what seems to be spontaneous combustion in the White House and Empire State Building. The "phasers" used in the *Star Trek* saga can heat, cut, explode, or dematerialize their targets. They can also knock living creatures unconsciousness, like a modern-day electric stun gun. "Freeze rays" also appear periodically, as in episodes of the TV series *Batman* (1966–1968) and *Underdog* (1964–1970). "By Any Other Name," a memorably absurd 1968 episode of the original *Star Trek* TV series, features a weapon that reduces its victims to plaster dodecahedrons (objects with twelve plane faces). A final class of "death rays" works so subtly that it leaves no mark on its victims. The hundreds felled by "blaster" fire in the original *Star Wars* movie trilogy (1977–1983) rarely show wounds, burn marks, or even holes in their clothing.

Regardless of *how* they affect their targets, the "death rays" depicted in popular culture take effect instantaneously. Unlike real-world lasers, they do not have to be held on target for any length of time; a split-second touch of the beam is sufficient. Fictional death rays are also much slower than real-world counterparts, which travel at the speed of light. They appear as discrete pulses of energy, traveling from gun to target slowly enough that human eyes can follow their progress. Would-be victims routinely escape death rays, seeing that they have been fired and *then* diving for cover.

Death-ray weapons in popular culture draw apparently limitless power from unspecified sources. Ship-based ones, like the Death Star's planet-killing beam in *Star Wars: A New Hope* (1977), are presumably powered by the same massive engines that drive the ships themselves. Hand weapons apparently draw their power from superbly engineered internal batteries. The batteries are small—*Star Trek*'s hand weapons are the size of a TV remote control, while those used in the TV series *Space 1999* (1975–1977) resemble stripped-down staple guns but store massive amounts of energy. The batteries are also very durable. Except in video games like *Doom*, they rarely become drained (demanding recharging or replacement) in the midst of battle.

Death ray weapons—even if called "lasers"—bear little resemblance to their real-world equivalents. They are far more closely related to the less-exotic guns depicted in popular culture, and their seemingly magical properties are the high-tech equivalent of cinematic six-shooters that can fire eight shots without being reloaded.

Related Entries: Action and Reaction, Law of; Flying Cars; Food Pills; Houses, Smart

FURTHER READING AND SOURCES CONSULTED

Franklin, H. Bruce. *War Stars: The Superweapon and the American Imagination*. Oxford University Press, 1990. Cultural history of the American obsession with "death rays" and other imagined, war-deciding weapons.

Garden, Tim. "High Energy Physics." *The Technology Trap*. Rev. ed. [draft]. 16 January 2001. <http://www.tgarden.demon.co.uk/writings/techtrap/draft5.html>. A critical, reasonably up-to-date discussion of lasers and other beam weapons for nonspecialists, used as background for this entry.

Hamilton, Mark. "Demystifying Lasers." Physics and Astronomy Outreach Program. University of British Columbia. 16 January 2001. <http://kepler.physics.ubc.ca~outreach/p420_96/mark/ham.html>. A brief, accessible general introduction to lasers, used as background to this entry.

Dinosaurs

Dinosaurs were the dominant group of land animals on Earth for nearly 150 million years, from the beginning of the Triassic period, 225 million years ago, to the Cretaceous period, 65 million years ago. First recognized and reconstructed by scientists in the 1840s, they have captivated the public ever since. Books on the giant beasts sell briskly, often stocked alongside displays of dinosaur T-shirts, models, jigsaw puzzles, and other paraphernalia. Discoveries of new dinosaur species are among the few types of science news guaranteed wide media attention. Mounted skeletons and life-sized models of dinosaurs are prominent in the collections of most major natural history museums and are consistently among their most-viewed exhibits. Dinosaur fossils have been "bones of contention" from the 1870s, when employees of rival paleontologists shot at each other, to the 1990s, when battles over a fossil *Tyrannosaurus rex* nicknamed "Sue" raged in court.

Interest in dinosaurs—among adults and especially among school-age children—has remained uniformly strong for decades. Scientists' understanding of dinosaurs, on the other hand, has changed substantially. The pre-1970s image of dinosaurs is so different from the post-1970s image that they might almost refer to different animals.

THE CHANGING IMAGE OF DINOSAURS

Dinosaurs are nominally reptiles, distantly related to modern-day crocodiles and lizards. From their discovery in the 1840s until the early 1970s, that knowledge dominated scientists' understanding of them. Reptiles are cold-blooded, drawing heat from their surroundings rather than generating it inside their bodies as birds and mammals do. They tend to move slowly, even sluggishly, and scientists long assumed that (because of their enormous size) the tendency was exaggerated in dinosaurs. The apparently small size of dinosaur brains combined this assumption of sluggishness with one of slow-wittedness. The standard pre-1970s image of the dinosaurs portrayed them as big and powerful but ponderous and dim-witted.

The modern era of dinosaur studies began around 1970, when research by paleontologists John Ostrom and Robert Bakker suggested that dinosaurs had been, in fact quick and highly mobile—perhaps even warmblooded. The idea was not new. The anatomical similarities between dinosaurs and birds had been noted, and a close evolutionary relationship between them had been suggested, since the mid-nineteenth century. Ostrom and Bakker gave the theory a stronger foundation, however, by closely analyzing what dinosaurs' anatomies implied about their lifestyles. Rather than plodding upright with its tail dragging, they argued, *Tyrannosaurus* darted like a roadrunner, head lowered and tail raised for balance. Large plant eaters like *Apatosaurus* were not confined to coastal waters, where the water supported their weight, but instead roamed the land, using their long necks to browse on leaves like elephants and giraffes do today. Water, when they did resort to it, may have served them as a radiator, dissipating heat from their enormous bodies.

A second major breakthrough in scientists' understanding of dinosaurs involved social behavior. Working in the badlands of Montana in the 1980s, John Horner and his colleagues discovered sites where dinosaurs had returned, year after year, to lay and hatch their eggs. The rich deposit of fossils surrounding the site, representing individual dinosaurs of both sexes and various ages, implied to Horner that dinosaurs lived in stable social groups and acted cooperatively to protect eggs, young, and perhaps aged or injured adults. Horner's conclusions reinforced those already advanced by Bakker, who argued that small, birdlike dinosaurs like *Velociraptor* had probably hunted in packs, cooperating to bring down larger prey.

Scientists' understanding of the dinosaurs' demise also changed in the 1980s. Physicist Luis Alvarez and colleagues argued in a landmark 1980 paper that a six-mile-wide asteroid had struck the Earth 65 million years ago, temporarily altering the climate and triggering the mass extinction that wiped out the dinosaurs, along with many other species. The Alvarez theory was backed by geological evidence of a 65-million-year-old impact somewhere on the Earth's surface, most famously a thin layer of sediment rich in iridium—an element rare on Earth but common in extraterrestrial objects. The iridium, Alvarez and his colleagues argued, had been part of the asteroid, which had vaporized on impact and become part of a globe-spanning dust cloud that altered the Earth's climate and triggered the mass extinction. The Alvarez theory swept away within a decade most of the hundred other theories that had been proposed in the preceding 120 years. The 1990 discovery of a twelve-mile-wide, 65-million-year-old crater at Chicxulub on the Yucatan Peninsula further strengthened it, by providing the "smoking gun" it had initially lacked.

OLD-STYLE DINOSAURS IN POPULAR CULTURE

The old image of dinosaurs as ponderous, dim-witted plodders casts a long shadow across popular culture. The menacing dinosaurs in Arthur Conan Doyle's novel *The Lost World* (1912), the original movie version of *King Kong* (1935), and time-travel stories like L. Sprague De Camp's "A Gun for Dinosaur" (1956) are created in this image. So too are the beasts featured in TV series like *The Flintstones* (1960–1966), comic strips like *Alley Oop* (1933–   ), and countless children's books like Syd Hoff's *Danny and the Dinosaur* (1958). The image also figures into many dinosaur-related jokes. One relatively sophisticated example, used by Gary Larson in a *Far Side* cartoon, shows a *Stegosaurus* addressing a roomful of other dinosaurs. The situation, he warns his audience, is grave: "The world's climates are changing, the mammals are taking over, and we all have brains the size of walnuts." The "Rite of Spring" segment of *Fantasia* (film 1940) depicts an old-style image of dinosaurs at their most majestic. Its animated characters transcend the limits of pre-computer-era special effects, and their ponderous movements have a kind of fluid grace. They are ultimately tragic figures, unable to escape or even comprehend the forces that destroy them.

Old-style dinosaurs work better as scenery than as individual characters. They are relegated to the margins and backgrounds of the stories in which they appear. The dinosaurs in *The Flintstones* and *Alley Oop*—even the Flintstone family's amiably brainless pet, Dino—are little more than props. The dinosaurs in *King Kong* or *One Million Years B.C.* (1940, 1966) are parts of long strings of obstacles that the heroes must overcome. Danny, not his lumbering pet, is the hero of *Danny and the Dinosaur*. Even the heroic-looking beasts in *Fantasia* are, in the end, part of the exotic landscape that serves as a backdrop for Igor Stravinsky's music. Their death is a natural process, like the eruption of volcanoes and the desiccation of the landscape.

Old-style dinosaurs come into their own, however, as symbols. The word "dinosaur" evokes, in everyday usage, not quickness and vigor but slow, plodding backwardness. To call something a "dinosaur" is to suggest that it is hopelessly behind the times, mired in the past and incapable of adapting to changing conditions. It was precisely these failings that, in the pre-1970s view of dinosaurs, led to their demise. The world changed, so the story goes, but the dinosaurs could not. Extinction was the price that nature imposed on them and, by implication, continues to impose on all species, individuals, and organizations that refuse to adapt. Mammals, in this view, are symbols of all the virtues that the dinosaurs supposedly lacked: quickness, flexibility, and adaptability. Their triumph is, implicitly, inevitable.

NEW-STYLE DINOSAURS IN POPULAR CULTURE

The new scientific understanding of dinosaurs that began to emerge in the mid-1970s gradually transformed dinosaurs' image in popular culture. The plot of Michael Crichton's *Jurassic Park* (novel 1990, film 1992) demands dinosaurs that are not just bigger and stronger than humans but also faster and more agile. The story's new-style *Tyrannosaurus* is capable not only of crushing cars and devouring their human passengers but of chasing the survivors when they try to escape in a jeep. Thundering down the roadway in hot pursuit, it is a more formidable threat than the slow-moving old-style *Tyrannosaurus* ever was. The greatest threat to the humans in *Jurassic Park*, however, comes not from *T. rex* but from the smaller, faster, and more intelligent *Velociraptor*. The human-sized raptors are able to penetrate buildings, opening or battering down doors in the process. They hunt in packs, attacking with such skill that in his last words big-game hunter Robert Muldoon expresses his professional admiration for the raptors that outflank and kill him. *Jurassic Park* could not have been written without the new view of dinosaurs that coalesced in the 1970s and '80s. Its thrills depend on dinosaurs that humans can neither outrun nor outmaneuver, and only sometimes outwit.

Not all new-style dinosaurs are depicted as vicious killers, however. Movies like *Dinosaur* (1999) and the popular *Land before Time* series (1988–) follow multispecies groups of dinosaurs who form surrogate families. Barney, the human-sized purple dinosaur of preschool children's television, is a classic new-style dinosaur: he sings, nimbly dances, and as a surrogate parent overflows with love for the children in his charge. Harry Harrison's "Eden" trilogy of science fiction novels (1984–1988) takes the idea of social dinosaurs a step farther. It takes place in an alternate history in which the dinosaurs did not become extinct but continued to evolve and founded a complex civilization in which still-primitive humans play a subordinate role.

Widespread acceptance of the idea that an extraterrestrial impact triggered the extinction of the dinosaurs has changed popular culture's attitude to their demise. Asteroid impacts strike down the well adapted and the ill adapted alike. Extinction from such a cause carries no moral stigma of having "failed to adapt." It makes the demise of the dinosaurs a lucky accident that *allowed* the mammals' rise to prominence, rather than the first act of a triumphant tale of progress.

Related Entries: Evolution; Extinction; Prehistoric Time

FURTHER READING AND SOURCES CONSULTED

Bakker, Robert T. *The Dinosaur Heresies*. Morrow, 1986. Horner, John. *Digging Dinosaurs*. Workman, 1988. Raup, David N. *The Nemesis Affair*. Norton, 1986. The ideas that revolutionized dinosaur studies in the 1970s and '80s, presented by scientists who helped to formulate them.

Farlow, James O., and M.K. Brett-Surnam, eds. *The Complete Dinosaur*. Indiana University Press, 1997. Overview of the current scientific views; more technical than Paul.

Lambert, David, and John Ostrom. *The Ultimate Dinosaur Book*. DK Publications, 1993. A visual "field guide" to the dinosaurs, with text and superb color illustrations covering fifty species from all major groups.

Mitchell, W.J.T. *The Last Dinosaur Book*. University of Chicago Press, 1998. Definitive history of the dinosaur's changing image in popular culture.

Paul, Gregory S., ed. *The Scientific American Book of Dinosaurs*. St. Martin's, 2000. A comprehensive collection of articles, written by specialists but designed for interested lay audiences.

Dolphins

Biologically speaking, dolphins are whales—small ones. They are marine mammals that breathe air, use sound to navigate and communicate, and swim using powerful strokes of their horizontal tails. Popular culture, however, treats dolphins differently than it does larger whales. Part of the reason may be size. Large whales, like the sperm or the blue, can grow as long as 100 feet and as heavy as 150 tons. They dwarf humans, just as humans dwarf bees and ants. Dolphins, like dogs and horses, are built to human scale. The bottlenose—the largest dolphin species and the only one to appear regularly in popular culture—seldom exceeds ten feet in length and 400 pounds in weight. The similarity in size allows humans and dolphins to interact in ways that would be implausible for humans and whales. In *The Deep Range* (1957), Arthur C. Clarke's science fiction novel about undersea farming, it is whales that are raised like cattle and dolphins that play the role of faithful sheepdogs.

Parallels with dogs are common in fictional treatments of dolphins. Fictional dolphins are not only intelligent enough to understand and respond to human commands but eager to please humans by doing so. They thus behave like the only "real" dolphins most people have ever seen: the ones performing in shows at aquariums and marine theme parks. Neither fictional dolphins nor their trained counterparts in the real world have obvious needs or interests of their own. Their lives revolve around, and are defined by, their human friends.

Flipper, the title character of a popular 1964–1968 television series, is a classic example. Flipper shares a deep bond with ten-year-old Bud Ricks and his teenage brother Sandy, sons of a park ranger stationed in the Florida Keys. The bond, however, is not between equals but between master and pet. Flipper lives in the equivalent of a doghouse: an enclosed pen alongside the Ricks family's dock. Playing with the boys and accompanying them on their adventures is, apparently, his greatest joy in life. He is the star of the program, but he seldom takes the initiative except to involve Bud in some new adventure or to summon help after Bud has gotten in (literally or figuratively) over his head. The parallels between *Flipper* and the legendary boy-and-his-dog series *Lassie* are too close to be accidental, and *Flipper*'s creators heightened them by ignoring or glossing over the substantial differences between a wild dolphin and a domesticated collie.

The 1973 film *Day of the Dolphin* places its two dolphin heroes in a different but still human-centered relationship. They are the officially the experimental subjects, but really the students, of scientist Jake Terrell, who through years of effort has taught them to speak English. The dolphins are presented as highly intelligent (the Bad Guys want to use them in a plot to assassinate the president of the United States), but their limited vocabularies and high-pitched "speaking" voices (provided by screenwriter Buck Henry) make them sound childlike. Their name for Terrell, "Pa," reflects their dependent relationship: he is both their teacher and their surrogate father. Far from minimizing this relationship, the film embraces it. The climax, in which Terrell sends the dolphins away in order to protect them, is heartrending (as intended), because we see the dolphins not as his partners but as his children.

Not all relationships between dolphins and humans are as deep and lasting as those in *Flipper* and *Day of the Dolphin*. Stories of sailors rescued from drowning or shark attack by benevolent dolphins have been part of maritime folklore for centuries. The dolphin in such stories appears suddenly and unexpectedly and, after ensuring the human's safety, goes on its way again. The dolphin-human interaction is fleeting rather than extended, but it is still defined by human, not cetacean, needs. Dolphins exist, once again, to serve humans.

Popular culture encourages us to see dolphins in comfortable, limited terms: trainable, loyal, and intelligent, like dogs. Scientists are far, however, from a complete understanding of dolphin intelligence and behavior. Dolphins may yet surprise us, as Douglas Adams suggested in the late 1970s in his radio play (and later novel) *The Hitchhiker's Guide to the Galaxy*. In Adams's farcical vision of the future, Earth's entire dolphin population takes leave of the planet just before Earth is vaporized as part of a galactic public-works project. Their parting words to the doomed humans who have treated them like pets: "So long, and thanks for all the fish!"

Related Entries: Intelligence, Animal; Whales

FURTHER READING

Cahill, Tim. *Dolphins*. National Geographic Society, 2000. Pictures from the society-produced IMAX film, and up-to-date text.

Carwardine, Mark, and Martin Camm. *DK Handbooks: Whales, Dolphins, and*

Porpoises. DK Publishing, 1995. Covers cetacean evolution, anatomy, physiology, social behavior, and communications; species-by-species illustrations.

Howard, Carol. *Dolphin Chronicles*. Bantam Doubleday Dell, 1996. A personal view of dolphins, emphasizing communication and the author's emotional bond with her research subjects.

Dreams

Dreams are images that occur in the mind during sleep. They occur most regularly and most vividly during REM sleep, a phase of the sleep cycle defined by increased brainwave activity and the rapid eye movement for which it is named. REM sleep, accompanied by dreams, occurs every ninety minutes or so. Humans between the ages of ten and the mid-sixties thus spend about 25 percent of an average night dreaming. There is no direct means of access to the content of those dreams. The best researchers can do is to wake the sleeper at the end of the dream and ask for a description.

The content of most dreams is made up of people and events from the dreamer's life, but these familiar elements are often juxtaposed in strange, disorienting ways. The world of dreams can, as a result, be a stressful place. Research suggests that unpleasant feelings outnumber pleasant ones two to one in dreams—a fact that may account for the relief that often accompanies a dreamer's return to the real world. The idea that dreams reveal the hidden, subconscious level of the dreamer's mind was introduced by Sigmund Freud around 1900, and became a pillar of twentieth-century psychology. Freud believed that dreams revealed unacknowledged desires kept suppressed during waking hours by the dreamer's conscious mind. Carl Jung concurred but disagreed with Freud's belief that the content of dreams was symbolic and required interpretation. Psychologists have suggested that dreams can also display the fruits of subconscious creativity. Samuel Taylor Coleridge's poem fragment *Kublai Khan* and August Kekule's formula for the molecular structure of benzene both made their first appearances in their creators' dreams.

Popular culture often treats dreams the way that Freud did, as coded expressions of the dreamer's deepest fears and desires. The musical *Oklahoma!*, which premiered on Broadway in 1943, includes a long scene in which Laurey, the heroine, imagines herself separated from Curly, the cowboy who loves her, by his brutish rival Jud Fry. The dream sequence is played as a ballet, which sets it off visually from the rest of the action. The film version of *Oklahoma!* (1955) stages the sequence on a highly stylized, eerily lit set that heightens the separation between dream world and real world. Films like *Spellbound* (1945) and *The Manchurian Candidate* (1962) burden their heroes with surreal dreams from which they must extract the clues to solve a mystery. Nicely-Nicely Johnson's dream, a turning point in the plot of the musical *Guys and Dolls* (stage 1950, film 1955), is also decidedly surreal. Johnson (who recounts the dream in a song) finds himself in a boat bound for heaven, but his passions for drink and dice put him in danger of falling overboard and being dragged below the surface by the devil. "Sit down!" his virtuous fellow passengers tell him, in the refrain that gives the song its title, "Sit Down, You're Rocking the Boat."

Most dreams in popular culture, however, are neither abstract nor symbolic. They are straightforward, realistic depictions of the dreamer's fears and desires. The beleaguered heroes of horror movies such as *Carrie* (1976), *Friday the 13th* (1980), and *An American Werewolf in London* (1981) suffer dreams in which, relaxing in seemingly safe places, they are attacked without warning by the forces of darkness. These scenes are visually identical to scenes set in the dreamer's real world—a convention that undermines the dreamer's, and the audience's, sense of security. Ambrose Bierce sustains this narrative device for an entire short story in "An Occurrence at Owl Creek Bridge" (1891). It begins with a man, condemned to hang, being pushed off a bridge with a noose around his neck. The rope breaks and, plunging into the river, the condemned man eludes his captors and escapes into the countryside. He makes his way home and has almost collapsed into his wife's arms when reality intrudes. His miraculous escape has all been a waking dream, played out in the split second before the rope jerks taut and breaks his neck.

Popular culture sometimes, in pursuit of an interesting story, goes well beyond science's understanding of dreams. The hero of Ursula K. Le-Guin's novel *The Lathe of Heaven* (1971) finds that his dreams literally come true. Freddy Krueger, the evil-incarnate villain of the *Nightmare on Elm Street* series of horror films (1984–1991), invades and manipulates the dreams of his young victims in order to scare them literally to death. The science-fantasy film *Dreamscape* (1984) has two psychic agents enter the troubled dreams of the president of the United States—one to assassinate him, the other to save him. The hero also, in subplots, uses his dream-entering talent for more routine tasks, such as trying to seduce his sleeping girlfriend. These stories play to audiences' experiences in their own dream-worlds, which feel as real to the dreamer as the external world that inspires them.

Related Entries: Mind Control; Psychic Powers

FURTHER READING

Flanagan, Owen J. *Dreaming Souls: Sleep, Dreams, and the Evolution of Mind*. Oxford University Press, 1999. Overview of modern research on sleep and dreams; organized around the question "What good is dreaming?"

Hobson, J. Allan. *Dreaming as Delirium: How the Brain Goes Out of Its Mind*. MIT Press, 1999. A clinical neuropsychologist's perspective on dreaming, tying dreams to chemical imbalances in the brain.

Jouvet, Michele. *The Paradox of Sleep: The Story of Dreaming*. Trans. Laurence Garey. MIT Press, 1999. Linked essays, surveying the state of the field, by one of its pioneers.

Earthquakes

Earth, whether as a planet or as a substance, is a traditional symbol of solidity. Ancient scholars—who saw Earth, Water, Air, and Fire as the four elements that composed and defined the physical properties of all things—believed that the presence of Earth was what made objects stable. The connection pervades everyday language: an impassive person is "stone-faced," and a blunt, uncomplicated one is "earthy"; irrevocable decisions are "carved in stone"; dependable things are "solid as a rock." People who remain steadfast in the face of adversity are likened metaphorically to rocks, as Jesus did of his disciple Peter ("On this rock I will build my church," Matthew 16:18). George and Ira Gershwin compared the stability of a lasting love affair to the stability of the giant Rock of Gibraltar in their 1938 song "Our Love Is Here to Stay." The central metaphor of "I Feel the Earth Move," Carole King's 1970 song about a passionate love affair, works *because* the earth (and Earth) stand for solidity. No force can be greater than one (passion, in the song) that makes them move.

Earthquakes are uniquely terrifying for just that reason. Floods, hurricanes, tornadoes, blizzards, and lightning may take lives and destroy property, but neither buildings nor people are supposed to last forever. "A generation comes and a generation goes," says the Bible (Ecclesiastes 1:4), "but the earth remains forever." Earthquakes are different: in the throes of one, the solidity and stability of the earth itself disappears. Everett Allen was being poetic when he titled his book on the great New England hurricane of 1938 *A Wind to Shake the World* (1976). In earthquakes, for minutes that survivors say pass like hours, the world really does shake.

Earthquakes happen because the crust of the earth—its hard outer shell—is not as stable as folklore implies. The crust is broken into huge, jagged-edged segments called plates, which lie atop a layer of molten rock (the mantle) thousands of miles deep. Driven by currents in the mantle, the plates move slowly but constantly over the surface of the earth, sometimes spreading apart, sometimes colliding, sometimes sliding one over the other, sometimes sliding past each other, sometimes locking together at their edges. The forces that drive the plates put enormous strains on them. Earthquakes happen when such strains build up until they are, in a catastrophic instant, released. They occur routinely where plates slide past each other (as in California) or where one is driven beneath another (as in Japan and Alaska). They can also occur, however, in areas far from any plate boundary. The quakes that devastated Lisbon, Portugal, in 1755 and Charleston, South Carolina, in 1886 fall into this category. So do the New Madrid earthquakes of 1811–1812, which changed the course of the Mississippi River and rang church bells as far away as New England.

Popular culture's depiction of earthquakes is shaped by memories of great cities devastated by quakes: San Francisco in 1906, Tokyo in 1923, Anchorage in 1964, Mexico City in 1985, and Kobe in 1995. Earthquakes in popular culture strike suddenly and without warning. They offer no hope of escape or even of last-minute preparations: all their victims can do is find shelter in doorways or under furniture and trust in the strength of the buildings they are in. Stories about earthquakes (unlike those about storms, fires, or floods, in which escape or preparation *is* possible) thus tend to *begin* with the disaster rather than ending with it. Movies like *Short Walk to Daylight* (made for TV, 1972), *Earthquake* (1974), and *Aftershock: Earthquake in New York* (made for TV, 1998) are less the stories of the quakes than the stories of city dwellers trying to make their way through the ruins of once-familiar landscapes. John Christopher's novel *The Ragged Edge* (1968) tells the same story on a grander scale, chronicling the disintegration of civilization amid a worldwide "epidemic" of earthquakes.

The most famous earthquake in popular culture is the hypothetical "Big One" that Californians believe will someday strike them. The "Big One" is occasionally depicted (*Richter 10*, by Arthur C. Clarke and Mike McQuay) or threatened (*Goodbye California*, by Alistair Maclean) in popular culture, but it exists mostly as a concept. It persists, as does the half-joking (but geologically absurd) idea that it will cause California to fall into the sea, because it allows Californians to cope with the less-powerful but still damaging earthquakes that do occur. Yesterday's earthquake is never the "Big One," which would have brought universal devastation and forced residents to confront the wisdom of living in a seismically active area. Diminished in this way, yesterday's earthquake can be tamed. It becomes, along with traffic jams and smog, part of "the price you pay for living here."

Related Entries: Comets; Lightning; Meteorites; Volcanoes

Earthquake damage in San Francisco, 1906. The quake, which killed 700, was the deadliest in U.S. history. Quakes of similar magnitude (8.3 on the Richter Scale) killed 25,000 in Mexico City in 1985 and 100,000 in Tokyo in 1923. Courtesy of the Library of Congress.

FURTHER READING AND SOURCES CONSULTED

Fradkin, Phillip L. *Magnitude 8*. University of California Press, 1999. The story of
    150 years of Californians' interactions with the San Andreas Fault, cov-
    ering both scientific and social issues.
Sieh, Kerry, and Simon LeVay. *Earth in Turmoil: Earthquakes, Volcanoes, and Their
    Impact on Humankind*. W.H. Freeman, 1998. A detailed, accessible survey
    covering risks from, predictions of, and preparations for, seismic activity.
USGS Earthquake Hazards Program. 10 June 2001. United States Geological Survey.
    12 June 2001. <http://earthquake.usgs.gov/>. Includes up-to-date maps,
    information on recent quakes, scientific background, a comprehensive
    FAQ, and links to regional sites.

Eclipses

The sun shines with a light of its own, and the moon with light reflected from the sun. An eclipse is the temporary blotting-out of light from one or the other. Caused by the relative motions of the sun, earth, and moon, they are both regular and predictable. Lunar eclipses, in which the shadow cast by the earth temporarily obscures the full moon, are beautiful but seldom gripping. They excite astronomers, but not the public. Solar eclipses—the darkening of the sun at midday—are an entirely different matter.

Solar eclipses occur when the moon moves directly between the sun and the earth. The sun is 400 times larger than the moon but also 400 times more distant from the earth, allowing the moon to blot out the sun exactly and cast a shadow on the earth's surface. Observers within the umbra, or heart of the shadow cast by the moon, see a total eclipse: the moon obscures the sun's bright disk for a few minutes, leaving only its atmosphere, or "corona," visible. The umbra, because of the earth's rotation, moves across the earth's surface, covering a strip about fifty miles wide. Observers on either side of that strip fall within the penumbra— the margins of the moon shadow—and see only a partial eclipse, in which part of the sun's disk is always visible. Solar eclipses are not rare; they occur, on average, once every eighteen months. The area in which any given eclipse will be seen as total covers one-half of 1 percent or less of the earth's surface. Eclipses can occur anywhere on Earth, but roughly 300 years will elapse between total eclipses in a given location. For most people, over most of human history, a total solar eclipse has been a once-in-a-lifetime event.

The idea that eclipses are routine and predictable has been part of Western science for centuries. It has not, however, completely displaced the much older idea that eclipses are the results of supernatural powers at work or signs that epochal events are unfolding. Because of this, and because of their rarity and dramatic appearance, total solar eclipses carry

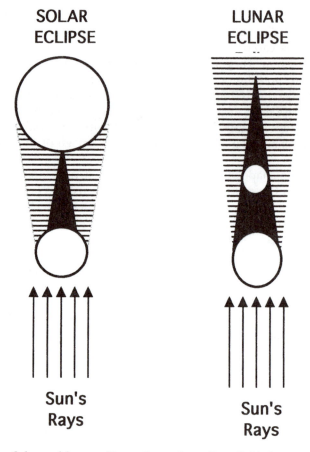

Solar and lunar eclipses. In a solar eclipse (left) the
moon casts a shadow on the earth: to observers within
the umbra (black) the eclipse is total; within the much
larger penumbra (dashed lines) it is partial. In a lunar
eclipse (right), the moon passes through the earth's
shadow; to observers on Earth, portions of the moon
that pass through the umbra (black) are blotted out. Illustration by the author, after diagrams in Richard P.
Brennan, *Dictionary of Scientific Literacy* (John Wiley,
1992), 189, 272.

more symbolic weight than most natural phenomena. The modern, scientific view and the premodern, mythical view of eclipses often coexist
in popular culture.

Hank Morgan, the hero of Mark Twain's *A Connecticut Yankee in King
Arthur's Court* (1889), is a man of 1878 who finds himself, memories

intact, in the year 528. Imprisoned by King Arthur and about to be burned at the stake, he remembers that a total solar eclipse is imminent and warns his captors that he will magically blot out the sun if they move against him. They scoff, the eclipse arrives on schedule, and Morgan claims responsibility. He secures his freedom and a job as Arthur's chief advisor. Equally important, he realizes that his nineteenth-century knowledge appears magical to sixth-century observers and that his newly won reputation as a magician gives him the social and political power to remake King Arthur's world in the image of the one he had left behind.

Eclipses also shape modern depictions of the Crucifixion. Three of the four Christian Gospels state that darkness covered the land on the day that Jesus was crucified; the most detailed states only that "the sun's light failed" (Luke 23:45, Revised Standard Version). Books like Jim Bishop's bestseller *The Day Christ Died* (1957) and epic movies like *King of Kings* (1961) and *The Greatest Story Ever Told* (1965) interpret those vague descriptions in fairly standard ways. The midday darkness is closer to dusk than midnight, and the sun is darkened or obscured rather than missing entirely. The effect is virtually identical to conditions at the height of a total solar eclipse, and audiences familiar with photographs and film footage of modern eclipses make the connection readily. The association of the Crucifixion with an eclipse likely persists because it is satisfying; the most ominous of natural events makes a suitable backdrop for the darkest day of the Christian year.

Observing solar eclipses without proper equipment is dangerous. Staring at one can permanently damage unprotected eyes; ordinary sunglasses and even welder's goggles do not offer adequate protection. News reports repeat these warnings whenever a solar eclipse is imminent. The reports seldom point out, however, that staring at the sun is dangerous under *any* circumstances and that eclipses are especially dangerous only because they tempt people to stare at the sun. The reports thus reinforce the widespread belief that the sun becomes uniquely dangerous during an eclipse—a modern, quasi-scientific echo of the ancient belief that powerful magic is at work when the sun darkens at midday.

FURTHER READING AND SOURCES CONSULTED

Brunier, Serge, and Jean-Pierre Luminet. *Glorious Eclipses: Their Past, Present, and Future.* Cambridge University Press, 2001. The history, science, and folklore of eclipses, extensively illustrated.

Littmann, Mark, et al. *Totality: Eclipses of the Sun.* 2nd ed. Oxford University Press, 1999. History, science, and folklore, plus viewing methods and safety tips.

"What Causes an Eclipse?" *Eclipse Tutorial.* 1998. Earthview. 4 May 2001. <http://www.earthview.com/tutorial/causes.htm>. Adapted from the book *Eclipses,* by Bryan Brewer. Source of the technical information in this entry.

Einstein, Albert

Albert Einstein (1879–1955) was central to the conceptual revolution that remade physics at the beginning of the twentieth century. Born in Germany, he graduated from a Swiss university and in 1901 took a job as a patent clerk in Berne. Pursuing theoretical physics in his spare time, he earned his Ph.D. in 1905 and in the same year published five papers in the *German Yearbook of Physics*. Three of the five papers were landmarks. One of them won Einstein the Nobel Prize for physics in 1921, and it is a measure of his genius that that paper was not the most significant of the three. That honor belongs to the paper dealing with the special theory of relativity: a complex working-out of the idea that motion can be measured only in relation to an arbitrarily chosen "frame of reference." The implications of special relativity were strange and unsettling. Objects moving at very high velocities increase in mass, diminish in length, and age more slowly. Mass and energy are interrelated; the famous equation that summarizes their relationship ($E = mc^2$) predicts that even tiny bits of mass can be converted into unimaginable quantities of energy.

The 1905 papers secured Einstein's reputation and eventually a teaching position in Berlin that allowed him to do science full time. He showed in 1915 that the principles of the special theory of relativity could be applied not only to reference frames moving at uniform velocities (his argument in 1905) but also to those undergoing acceleration. Experimental confirmation of this "general theory of relativity" made Einstein world famous. It also set the stage for the later decades of his career, when his social activism won him more recognition than his fruitless search for a "unified field theory." Einstein's principal social cause was peace. Ironically, his greatest impact on world affairs stemmed from a 1939 letter, composed by fellow scientists and reluctantly signed by Einstein, urging President Franklin Roosevelt to develop an atomic bomb. He later regretted the letter and, from 1945 until his death a decade later, he campaigned tirelessly against the nuclear weapons that, he believed, had supplanted Nazi Germany as the principal threat to world peace. The Einstein of popular culture is, on closer examination, actually two different Einsteins. The first is a wizard whose ideas are incomprehensible to mere mortals. The second is an ordinary man with a full complement of ordinary quirks.

Einstein-the-wizard inspired countless half-serious jests to the effect that only five (or ten or twenty) people in the world could understand the theory of relativity. The same sentiment inspired an anonymous limerick popular in the 1930s that linked him with avant-garde writer Gertrude Stein and sculptor Jacob Epstein:

> I don't like the family Stein:
> There's Gert and there's Ep and there's Ein.
> Gert's poems are bunk,
> Ep's statues are junk,
> And nobody understands Ein.

The equation $E = mc^2$ is an integral part of the Einstein-the-wizard image. Most people with a high school education can recognize it and associate it with Einstein. It functions, in virtually any image-driven branch of popular culture, as visual shorthand for "Very smart people are probing the mysteries of the universe here." Few nonscientists can readily define its terms or explain its significance, which makes it an even more powerful symbol. Like a wizard's magic spell, its power can be unleashed only by a select few.

Einstein-the-ordinary-man appears in three or four standard black-and-white images on posters, mugs, bookmarks, and other merchandise. The tools of his trade are absent. Instead, he rides a bike, makes a face, or simply looks benignly out of the picture. Significantly, all these images are from Einstein's later years. He looks far less like the intense young man of the 1910s than like somebody's rumpled but beloved uncle. In the film *I.Q.* (1993), Einstein really *is* the uncle of the fictitious Catherine Ryan (Meg Ryan), whose romance with a brilliant but uneducated garage mechanic he does everything possible to promote. Walter Matthau plays Einstein as a kindly old man who loves his niece, enjoys the company of his friends, and (almost as an afterthought) is a brilliant theoretical physicist. The most familiar anecdotes about Einstein underline his ordinary-person status: his tendency to forget appointments, for example, and his lackluster record in school.

Popular culture's two images of Einstein complement each other. Einstein-the-wizard, unraveling the secrets of the universe and surveying the road to the atomic bomb, could be a dark, disturbing figure. The strong presence of Einstein-the-ordinary-man humanizes him, however, and renders him unthreatening. Mad scientists don't ride bicycles, make funny faces, or play matchmaker for their nieces. On the other hand, the apparent simplicity of Einstein-the-ordinary-man makes the transcendent insight of Einstein-the-wizard seem all the more striking and even unfathomable.

Related Entry: Relativity

FURTHER READING

Clark, Ronald W. *Einstein: The Life and Times*. World Publishing, 1971. Einstein's life and work, recounted by a noted biographer.

Einstein, Albert. *Ideas and Opinions*. (1954; reprint Bonanza Books, 1988). Einstein's worldview, presented in his speeches and essays.

Hoffman, Banesh, and Helen Dukas. *Albert Einstein: Creator and Rebel* (1972; reprint New American Library, 1989). Einstein's life and work, recounted by two former colleagues.

Electricity

Electricity is the blanket term for phenomena caused by the movement of electrons and the charges they carry. "Static electricity" is an excess of charge that can be built up in certain materials by rubbing them with other materials. "Current electricity" is a stream of electrons, produced by chemical reactions (like those in batteries) or by the movement of a wire through a magnetic field. Static electricity, known since ancient times, has remained little more than a curiosity. Current electricity, first investigated in the late eighteenth century, soon transformed the world. Knowledge of how to generate, transmit, and use it accumulated rapidly, triggering the "Second Industrial Revolution" in the last third of the nineteenth century.

Electricity was the heart and soul of the Second Industrial Revolution. Electric motors made possible the motorized streetcar, the subway, the elevator, and—through them—the modern city of high-rises ringed by residential suburbs. High-capacity generators enabled the electric lighting of public spaces, city streets, and (beginning in the 1880s) private residences. The wiring of the industrialized world began in the cities and spread steadily outward over the next fifty years. Electric light was the first and greatest attraction for newly wired households, but other electric appliances followed it: fans, irons, washing machines, refrigerators, and (by the 1940s) air conditioners. The everyday routines of life in the industrialized world became and have remained inseparably interwoven with electricity.

Popular culture depicts electricity less as a natural phenomenon than as the lifeblood of modern civilization. Electricity, like radiation, is accorded powers more supernatural than natural: it can strike down the living, animate the dead, or contravene the laws of nature. Characters who can reliably harness its powers know things that few mortals know and can do things that few mortals can. These attitudes crystallized in Mary Shelley's 1819 novel *Frankenstein*, written when experiments with

Laying the Electrical Tube.

Laying electrical cables in New York, drawn by W.P. Snyder for the 28 June 1882 edition of *Harper's Weekly*. The enormous costs involved in "wiring" an area meant that customer-rich cities enjoyed electrical service long before isolated rural areas did. Courtesy of the Library of Congress.

current electricity were still new. Her hero, Victor Frankenstein, is a se-
cretive figure whose gothic castle-laboratory recalls a wizard's inner
sanctum. He uses electricity to animate the patchwork creature he
stitches together—grasping for the godlike power to bestow life at will.

The popular culture of the Second Industrial Revolution takes a more
optimistic view of electricity. Electricity still confers near-magical powers
on the few who can understand and manipulate it, but these people are
more likely to use their power to benefit society than to gratify themselves. Rudyard Kipling pays tribute in his 1907 poem "The Sons of
Martha" to heroic electrical workers who "finger death at their gloves'
end as they piece and repiece the living wires." Jules Verne's heroes have
many of their adventures in electric-powered vehicles, as does Tom
Swift, hero of a turn-of-the-century series of boys' adventure novels.
"Long Tom" Roberts—assistant to Doc Savage in a series of pulp-
magazine adventures from the 1930s and '40s—is physically frail but a
worthy ally because of his mastery of electricity. Roberts is frequently
dubbed a "wizard," as were real turn-of-the-century electricity experts
like Thomas Edison, Charles Steinmetz, and Nikola Tesla.

By the second half of the twentieth century, electricity was neither a
scientific nor a technological novelty. It appears in popular culture as
part of the fabric of everyday life—requiring attention or comment only
when it stops flowing. People who use their specialized knowledge to
keep it flowing still appear as heroic, wizard-like figures, and not only in
obvious places like power-company commercials. The nameless title
character of Jimmy Webb's song "Wichita Lineman" (1968) endures long
absences from his own home in order to safeguard the lines that carry
power to thousands of others. The power-company workers in a mid-
1970s Miller beer commercial ("And now comes Miller time. . . .") are
heroic because they hold in their hands "enough power to light up the
county."

The *loss* of electrical power usually signifies disaster in late-twentieth-
century popular culture. When a power failure plunged the Northeast
into darkness in the fall of 1965, many residents initially feared a nuclear
attack by the Soviet Union. "The Monsters Are Due on Maple Street," a
1960 episode of the TV series *The Twilight Zone*, tracks the rapid disintegration of civility and sanity in a suburban neighborhood whose electrical power is (unknown to its residents) being manipulated by aliens.
When Nantucket Island and its residents are cast back in time to the year
1250 B.C. in S.M. Stirling's *Island in the Sea of Time* (1998), their need to
ration electricity is acute, and their sense of loss palpable. Larry Niven
and Jerry Pournelle's novel *Lucifer's Hammer* (1977), which concerns the
catastrophic impact of a comet on Earth, ends with a group of survivors

anticipating the revival of a dormant power plant. Reestablishing civilization depends, for them, on reestablishing the flow of electricity.
Related Entries: Lightning; Magnetism; Radiation

FURTHER READING

Hughes, Thomas P. *Networks of Power: Electrification in Western Society, 1880–1930.* Johns Hopkins University Press, 1988.

Nye, David E. *Electrifying America: The Social Meanings of a New Technology, 1880–1940.* MIT Press, 1990. Two authoritative histories of the immense and wide-ranging impact of electricity on everyday life.

Ryan, Charles William. *Basic Electricity: A Self-Teaching Guide.* 2nd ed. John Wiley, 1986. A comprehensive, workbook-style introduction that proceeds in careful steps from absolute basics through circuit design.

Elephants

Elephants are the largest living land mammals, and the largest of all living mammals after the whales. Two species survive: the larger African elephant and the smaller Indian elephant. Fully grown African males can reach heights of thirteen feet at the shoulder, and weights of six to eight tons; even newborn calves are formidable, standing three feet high and weighing 200 pounds. The elephant's most distinctive features—its ears, trunk, and (in males) tusks—are all highly functional. The ears act as radiators, the tusks as weapons and digging tools. The trunk functions as a kind of hand, conveying food (mostly leaves and grasses) to the elephant's mouth and drawing up water for drinking or bathing. Muscular "fingers" at its tip give it the ability to manipulate even small objects with great precision. Elephants' strength, dexterity, and high intelligence make them valuable to humans; they have been domesticated and trained since ancient times as beasts of burden, cavalry mounts, and performers.

Elephants reach maturity at twenty-five but begin mating as early as eight or ten, with females producing a calf about every four years. They live an average of sixty years in the wild, but may reach eighty under ideal conditions. Their low birth rates and long lifetimes create herds that are small, tightly knit, and organized by sex and age. A typical herd, invariably led by a large male, might consist of ten to fifteen females and their young under the immediate leadership of an older female. Younger males typically remain on the fringes of the herd as scouts and guards, but the entire herd cooperates to protect vulnerable members and raise orphaned calves.

Elephants function in popular culture in two distinct ways: as emblems of the exotic and as models of behavior that humans find exemplary. In the first role, elephants are ubiquitous. In the second, they are confined largely to children's stories, but they account for some of the most iconic characters in all of children's literature.

Elephants seem exotic to Western audiences because, like the giraffe, rhinoceros, and hippopotamus, they are so unfamiliar. Zebras look like horses, and antelopes like deer, but no other creature alive resembles an elephant. Popular culture's stock references to elephants hinge on this exotic uniqueness. Most of the vast repertoire of "elephant jokes," for example, derive their humor from the beasts' unusual size (Q: How can you tell that there are elephants in your refrigerator? A: They leave footprints in the butter). The widely used expression "white elephant" implies an unwanted object that is not just useless but bizarre.

The elephants of children's literature are kind and caring, just as the monkeys are fun loving and the rhinoceroses bad tempered. They act as humans believe that they themselves should, forming close-knit communities in which each member looks out for the welfare of the others. "The Elephant's Child," one of Rudyard Kipling's *Just So Stories* (1902), is the story of a much-spanked young elephant who leaves his multispecies "family" but in time returns to it and is accepted as an equal. *Dumbo* (film, 1942) is the story of an unusual young elephant's search for acceptance. The title character of *The Saggy Baggy Elephant* (1947), one of the best-selling "Little Golden Books" of all time, is in search of both a family and an identity At the climax, he is rescued (and adopted) by the first *other* elephants he has ever seen. All three stories reach happy endings when the temporarily isolated hero is welcomed into a close-knit community.

Not all elephant tales deal with isolation, however. The adventures of Laurent de Brunhoff's *Babar* (1935–) take place within a jungle society built and ruled by civilized elephants. The preservation of the group—against a tribe of conquest-minded rhinoceroses, for example—is again central. David McKee's *Elmer* series (1989–) features a brightly colored elephant whose "job" is to entertain his gray-skinned brethren with practical jokes and celebrations. Dr. Seuss's famous character Horton takes the elephant ethic of concern for others beyond the bounds of his species. In *Horton Hatches the Egg* (1940) he looks after the offspring of an irresponsible bird, and in *Horton Hears a Who* (1954) he saves an entire microscopic race from destruction.

Horton's famous catch-phrases—"An elephant's faithful, one hundred percent," and "A person's a person, no matter how small"—are deeply humane. So too is Babar's determination to preserve civilization, and Elmer's to do his bit for the herd. Admirable in an elephant, these behaviors would be equally admirable in a human (the behavior of H.A. Rey's "Curious George," by contrast, is endearing in a monkey but would be insufferable in a human). Fictional elephants, more than any other fictional "jungle animal," are human surrogates. Their concern for family, community, and each other makes them what the rest of us aspire to be.

Related Entries: Chimpanzees; Gorillas; Intelligence, Animal

FURTHER READING

The Elephant Information Repository. <http://elephant.elehost.com>. A full discussion of elephant biology, behavior, and conservation.

Granig, Karl, and Morton Saller. *Elephants: A Cultural and Natural History*. Koneman, 1999. Large-format coverage of biology, folklore, and interactions with humans.

Moss, Cynthia. *Elephant Memories: Thirteen Years in the Life of an Elephant Family*. University of Chicago Press, 2000. A landmark report, originally published in 1988, on the author's close observation of elephant behavior in southern Kenya.

Epidemics

Epidemics are outbreaks of disease—especially communicable disease—that affect large numbers of people. They cut through human populations like wildfire through a forest: most destructive where their human "fuel" is most densely packed, and prone to burn out in areas where fresh "fuel" is scarce. They also spread like wildfires, except following routes of road, rail, and sea travel instead of wind currents. Epidemics often leap from one major population center to another, carried unwittingly by infected travelers. Epidemic diseases spread from one victim to another in a variety of ways: by air (influenza), tainted water (cholera), tainted bodily fluids (AIDS), or contact with infected animals (malaria). Those closest to the victims—family members, close friends, and medical practitioners—are especially likely to become victims themselves.

Epidemics reach beyond cities, but they thrive in urban settings. Densely packed urban populations provide a rich supply of victims, converging transportation routes encourage spreading, and sanitation is often inadequate. Periodic epidemics of cholera, yellow fever, malaria, and smallpox were facts of life, well into the nineteenth century, in cities as modern as London and New York. Advances in medicine and public health have suppressed these diseases in the developed world, but only smallpox has been eradicated entirely. Epidemics of incurable diseases remain a potent threat, as the AIDS crisis makes clear.

The death tolls of major epidemics can be staggering. The influenza epidemic of 1918–1919 killed 20 million worldwide, more than the just-concluded First World War. Epidemics also wreak havoc in other ways. They often cut down entire families, especially in societies where the families (rather than health-care workers) are expected to nurse the sick. They also gut population centers; the medieval "Black Death" killed 30 to 40 percent of Europe's population, but death tolls in individual towns and villages reached 80, 90, or even 100 percent. These patterns of con-

Treating the influenza epidemic of 1918–1919. Masked to prevent infection, Red Cross nurses demonstrate patient care techniques at an emergency aid station in Washington, D.C. Courtesy of the Library of Congress.

centrated death often cause the temporary collapse of political and social institutions.

Uncontrolled epidemics typically cause massive, temporary disruptions of everyday life rather than permanent social collapse. Control, though sometimes aided by vaccines and curative drugs, usually depends on public health and public education campaigns designed to change the way members of the society live. Popular culture's treatment of epidemics tends, however, abandon this middle ground for one of the two extremes. Epidemics either succumb to a "magic bullet" developed by heroic scientists or prove so intractable that they destroy civilization and usher in a new Dark Age.

The epidemics in the heroic-scientist stories stand at the center of the story: high-stakes puzzles that the heroes solve by the end of the story. Patients—often including one of the heroes—die, but the surviving heroes solve the puzzle in time to avert a widespread catastrophe. The title character in Sinclair Lewis's novel *Arrowsmith* (1925) saves hundreds in the present and, implicitly, thousands in the future by curing a tropical disease. The self-described "virus cowboys" in the film *Outbreak* (1995) isolate (and create a cure for) a deadly government-engineered disease while it is still confined to a single California town. The scientists in Michael Crichton's *The Andromeda Strain* (novel 1969, film 1971) uncover the secret of an extraterrestrial virus just in time to keep it from spreading across the continent. A medical solution is, nearly always, sufficient to stop the epidemic quickly and cleanly. The heroic scientists rarely face the hard, thankless work of changing the public's behavior.

The epidemics in the end-of-the-civilization stories often take place offstage, setting the story in motion. The story itself focuses not on the epidemic but on the heroes trying to come to terms with the world into which it has plunged them. Coming to terms is often equated not with rebuilding the old world but with abandoning the last vestiges of it in favor of a different way of life—if not forever, then for many lifetimes. George R. Stewart's novel *Earth Abides* (1949) traces the rapid erosion of urban civilization and ends with the hero's final acceptance of the Stone-Age lifestyle his postepidemic children will lead. Stephen King's *The Stand* (1982, revised 1990) features two groups of survivors wandering in an America devastated by biological warfare research gone bad. The forces of good—called by visions and bound by faith—rally behind a saintly, religious older woman; the forces of evil cling to the materialistic, technological preplague world, clinging even to a stray atomic bomb.

Both types of stories are rooted in a deep-seated faith in the power of science and technology to improve our lives. Stories in the first group reaffirm that faith. Stories in the second group spin out a nightmare in which that faith is revealed, too late, to be hollow and worthless.

Related Entries: Experiments; Experiments on Self; Miracle Drugs

FURTHER READING AND SOURCES CONSULTED

Garrett, Laurie. *The Coming Plague*. Farrar, Strauss and Giroux, 1994. Detailed, journalistic treatment of emerging epidemic diseases.

McNeill, William H. *Plagues and Peoples*. Anchor, 1976. Classic historical survey of the impact of epidemics on human history; source of the historical data cited in this essay.

Shilts, Randy. *And the Band Played On*. St. Martin's, 1987. The early years of the AIDS epidemic from the perspectives of both scientists and victims.

Evolution

Evolution is the formation of new species from existing species. Charles Darwin (1809–1882) was not the first to propose the idea, but he was, in his 1859 book *On the Origin of Species*, the first to marshal a wide range of data to support it. Darwin's book generated much scientific support for evolution but little for the mechanism—natural selection—that he believed lay behind it. Full scientific acceptance of Darwin's theory came only in the 1930s, when it was interwoven with the then-new science of genetics in what became known as the "modern synthesis." This synthesis remains, seven decades later, the standard scientific view of how evolution works.

Evolution in the Darwinian model is the result of what he called "natural selection." The individual organisms that make up a species are not identical. Each new generation will, because of random genetic mutations, include a few individuals with traits that improve their chances of survival and a few with traits that diminish their chances. Members of the latter group tend to die young, leaving few if any offspring to carry their genes. Members of the former group tend to thrive, breeding more prolifically than "average" members of the species and leaving many offspring to carry their genes. Individuals with beneficial traits survive more readily, live longer, and produce more offspring than their "average" species mates. Over many generations, therefore, their descendents (which carry the beneficial trait) gradually come to outnumber the descendents of "average" individuals (which lack the trait). The steady accumulation of beneficial traits can, over millions of years, produce ancestors and descendants different enough to be considered members of separate species.

Evolution, in the Darwinian model, is an ongoing process of adaptation to environmental conditions. Those conditions, and the definition of "beneficial," vary from place to place. A trait that is beneficial in one local environment may be useless or even detrimental in another. Separate populations of a single species, living in different areas with different environmental conditions, may evolve in very different directions, giving multiple descendent species a common ancestor. Conditions, and the definition of "beneficial," vary also over time. A species adapted in highly specialized ways to particular environmental conditions may face extinction if those conditions change too much, too fast.

Darwinian evolution is, as a result, resolutely nondirectional. It sees the history of life not in terms of steady progress toward a specific goal but in terms of adaptation—always temporary—to the conditions of the moment. It is not a random process (beneficial traits tend to accumulate, and detrimental ones tend to vanish quickly), but it is driven by the randomness of genetic mutations and environmental changes. "There is," Darwin wrote in *Origin of Species*, "a grandeur in this view of life," and many working biologists concur. Some, like Stephen Jay Gould and Richard Dawkins, have written eloquently about it. The general public, however, favors a different view of evolution, one proposed by Darwin's predecessor Jean-Baptiste Lamarck (1744–1829) and widely accepted by scientists before the advent of the modern synthesis.

The Lamarckian model portrays evolution as linear and orderly, driven by an innate tendency of organisms to become more complex from generation to generation. The history of life is, in the Lamarckian model, a story of steady upward progress in which crude, simple organisms give way to sophisticated, complex ones. Humans are not simply one species among many but the topmost rung on the ladder of evolutionary progress. Popular culture reflects and reinforces the general public's preference for the Lamarckian over the Darwinian model of evolution. Except in works specifically intended to explain the Darwinian model, Lamarckian images and metaphors predominate.

The language used in popular culture to describe the process of evolution has progressive overtones that are central to the Lamarckian model but alien to the Darwinian. Descriptions of now-extinct lineages as evolutionary "dead ends" or "blind alleys" are common. They imply the existence of an evolutionary "main line" from which it is unwise to stray. Statements that the dinosaurs (or dodo birds or Neanderthals) became extinct because they "failed to evolve" are equally common, and equally Lamarckian. They imply that evolution is driven by internal, not external, forces and that extinction is nature's penalty for the extinct species' absence of ambition or will. They are also, like many of the Lamarckian model's implications, easy to connect to cherished cultural values like enterprise, adaptability, and ingenuity. Lamarckian evolution lends itself to morality tales in a way that Darwinian evolution does not.

The history of life is often represented in popular culture as a series of "ages," each named for the class of animals that, by implication, dominated it. The Age of Fishes is succeeded, in countless popular science

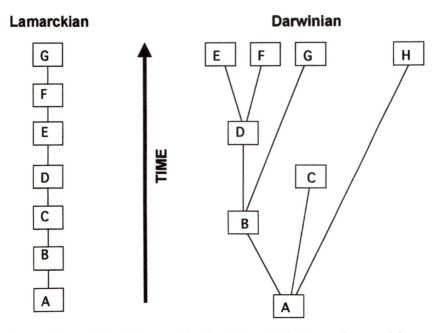

Lamarckian and Darwinian models of evolution. In the Lamarckian model (left) every lineage follows the same ladderlike path. Simple organisms (species A) steadily evolve into increasingly complex ones (species B through G). In the Darwinian model (right) every lineage follows a unique, bushlike path. A given species (such as A) may have multiple descendent species: some that change significantly (B), some that remain stable (H), and some that become extinct (C). The unique shape of each "bush" is created by the random mutations and environmental changes that drive the Darwinian model.

books, by the Ages of Amphibians, Reptiles, Mammals, and Man. The visual equivalent is a vertical column divided by horizontal lines into geological eras, periods, and epochs—each of which is embellished with a small drawing of the "dominant" animal. The animals chosen as illustrations invariably represent the familiar sequence of vertebrate classes listed above. Where several members of the same class—several mammals, say—are used to represent successive subdivisions of a single age, the largest species are reserved for the most recent times. Both conventions reflect and reinforce Lamarckian attitudes. The animals depicted get steadily more like us as time goes by: invertebrate gives way to vertebrate, marine to terrestrial, cold-blooded to warm-blooded, four-legged to two-legged. Thus, from our human-centered perspective, they get steadily closer to the "goal" of the evolutionary process: *Homo sapiens*.

The Lamarckian view of evolution as linear and progressive is particularly evident in portrayals of human evolution. Several species of the genus *Australopithecus* and several more of the genus *Homo* constitute our immediate evolutionary "family": the hominids. Theories about which of them were our direct ancestors vary widely and change frequently. The graphics that represent these theories in the popular press, however, invariably show the lineage leading to *Homo sapiens* as a straight line in the center of the frame. All other lineages branch from that main line like railway sidings, stopping short of the top of the page so as to represent extinction. This "family tree" diagram is superficially Darwinian, but its most striking visual element is Lamarckian: the long, straight, upward line of progress that leads to us.

The aesthetic and emotional appeal of the Lamarckian model of evolution is immense. The model implies that nature has direction and purpose, and it leaves ample room for an Intelligent Designer to supply a greater purpose. The Lamarckian model fails, however, to match the Darwinian model's ability to bring an elegant, unifying order to data from fields as disparate as genetics, paleontology, and taxonomy. Among working scientists, this ability gives Darwinian evolution an aesthetic and emotional appeal of its own—the "grandeur" of which Darwin wrote. Conveying that sense of grandeur to the general public is a challenge, one that as yet scientists have tackled with only limited success.

Related Entries: Darwin, Charles; Evolution, Convergent; Evolution, Human; Extinction; Genes; Mutations

FURTHER READING

Bowler, Peter J. *Evolution: The History of an Idea.* Rev. ed. University of California Press, 1989. The best historical survey of theories of evolution from the eighteenth century.

Dawkins, Richard. *River Out of Eden.* Basic Books, 1996. Introductory survey of one major strand in modern evolutionary theory: genes as the central agents in evolution.

Gould, Stephen Jay. *Full House: The Spread of Excellence from Plato to Darwin.* Harmony Books, 1996. Detailed argument for evolution as nondirectional and nonprogressive, and critique of popular culture's Lamarckian leanings.

Weiner, Jonathan. *The Beak of the Finch: A Story of Evolution in Our Time.* Vintage Books, 1995. How modern studies of the finches of the Galapagos Islands, one of Darwin's inspirations, support and extend his theories.

Evolution involves the development, over many generations, of new traits in existing species of organisms. New traits, according to the Darwinian view of evolution that most scientists hold, make their initial appearance as a result of random, natural genetic mutations. Beneficial traits, which improve an individual's chances of surviving, are statistically more likely to be passed on to offspring. Over time, therefore, they become more and more common within the species. The gradual accretion of such new traits over vast spans of time eventually produces organisms so different from their direct ancestors that they constitute a different species.

Convergent evolution is the appearance of similar structures in species that have no direct evolutionary relationship. It happens when similar mutations, and similar environmental pressures, produce similar adaptations in otherwise unrelated species. Wings are a classic example of convergent evolution at work. Both vertebrates and insects have them, and both use them to fly, but vertebrate wings and insect wings are "built" differently, develop differently, and function differently. Features produced by convergent evolution are said to be "analogous": similar in function and superficial appearance but different in structure. "Homologous" features, like human hands and whale flippers, are something else: they look similar but also have similar structures, because their "owners" share a common ancestor. Convergent evolution can also produce species that, though geographically and genetically distant from one another, look similar because they evolved to fit similar environmental conditions. Ostriches and emus—large, flightless birds that live in Africa and Australia, respectively—are a familiar example of this rare phenomenon.

Popular culture invokes convergent evolution on a far grander scale than most biologists would dream of, or accept. Convergent evolution in the real world involves body parts or, rarely, entire organisms. Convergent evolution in the fictional worlds of popular culture involves entire planetary ecosystems. It is used, usually in a sentence or two of throwaway explanation, to justify the otherwise inexplicable tendency of intelligent alien species to be humanoid.

Popular culture's use of convergent evolution makes some sense at a superficial level. The aliens encountered by human space travelers nearly always inhabit worlds that are Earthlike in size, gravity, climate, and topography. These similarities suggest that the environmental pressures that shaped the evolution of the alien organisms would be similar to those that shaped the evolution of life on Earth. If convergent evolution could produce the similar-looking ostrich and the emu for ecologically similar niches on different continents, why couldn't it produce similar-looking (that is, humanoid) species on ecologically similar planets?

The problem with such an argument (and with popular culture's use of convergent evolution) is that evolution is driven by genetic as well as environmental pressures. Even the most distantly related vertebrates on Earth have substantially similar DNA. There is no reason to believe that alien species would have DNA, cell structure, or physiology remotely similar to *any* Earth-born species. There is not *necessarily* any reason to believe that they would even have a biochemistry based on carbon. Even if they evolved in identical environmental settings, organisms whose genetic and biochemical building blocks were profoundly different would be unlikely to look anything like each other. The creators of *Star Trek*, responsible for more humanoid aliens than anyone else in popular culture, finally addressed this issue in "The Chase," a 1994 episode of *Star Trek: The Next Generation*. The crew of the *Enterprise* discovers that long ago an alien species traveled the galaxy "seeding" environmentally promising worlds with DNA. Humans, Klingons, Vulcans, and the rest of *Star Trek*'s humanoid species are all products of that experiment: distant relatives after all.

Popular culture's fondness for convergent evolution and a universe full of humanoid aliens owes more to practical necessity than scientific reality. It is least common in books, where alien species and alien worlds exist only in the author's descriptions and the readers' imaginations. Free of the burdens of actually creating replicas of what they imagine, writers can make their aliens look any way they want. Movie and TV producers, on the other hand, are obliged to balance scientific plausibility against such practical constraints as budgets and shooting schedules. Particularly in TV, where both budgets and schedules are generally tighter, scientific plausibility tends to lose. Earthlike planets allow location shooting to replace elaborate sets, and humanoid aliens can be played by human actors in exotic costumes and makeup. Humanoid aliens can also "act" using familiar human facial expressions, a boon to both actors and directors. It is no accident that alien characters' prosthetic makeup routinely alters their noses, ears, and foreheads but almost never obstructs the most expressive parts of the face: the eyes, mouth, and lower jaw.

Related Entries: Cloning; Evolution; Life, Extraterrestrial; UFOs

FURTHER READING

Andreadis, Athena. *To Seek Out New Life: The Biology of Star Trek*. Three Rivers, 1999. Jenkins, Robert, and Susan C. Jenkins. *Life Signs: The Biology of Star Trek*. HarperCollins, 1999. Useful discussions of popular culture's handling of alien life forms, using *Star Trek* as a source of examples.

Emmerich, Kevin, and Laura Cunningham. "Convergent Evolution." *Desert Lizard Ecology in Death Valley National Park*. 17 August 2001. <http://cluster4.biosci.utexas.edu/deathvalley/converge.htm>. Visually awkward but content-rich discussion of convergent evolution in the real world, with illustrations.

Evolution, Human

Scientists have in the last half-century formulated countless theories of human evolution. Virtually all are variations on a set of ideas that have remained stable for the same period or more. This durable set of ideas constitutes the basic scientific view of how our species, *Homo sapiens*, came to be. The view of human evolution presented in popular culture has also remained stable for a half-century or more. It rests, however, on a very different set of assumptions. Both views agree that humans evolved as other species do but diverge from each other sharply after that.

The scientific consensus on human evolution rests on the idea that humans share a common ancestor with the great apes. We look different than the apes (gorillas, orangutans, and chimpanzees) because our lineage diverged from theirs between six and ten million years ago. The hominid lineage itself split several times after that, producing several species of which ours (*Homo sapiens*) is the only survivor. *Homo sapiens* acquired its distinctively "human" traits—large brains, flat faces, flexible hands, multipurpose teeth, and an upright posture—because hominids born with one or more of those traits survived more readily and passed on their genes to more offspring than hominids born without them. Our bodies are, as a result, a patchwork of features inherited from distant ancestors and features recently evolved. The drawbacks of the former, such as quadraped-style knees that are prone to injury when we walk upright, are outweighed by the benefits of the latter, such as flexible hands that are always free to manipulate objects.

Popular culture's view of human evolution proceeds from the idea that we evolved *from* the apes. Our distant ancestors, according to popular culture, looked like an amalgam of gorillas and chimpanzees. A straight evolutionary line connects them to us, leading upward like a ladder. Most fossil hominids occupy intermediate "rungs" on this ladder, standing somewhere between apes and modern humans. *Homo sapiens* acquired its distinctive combination of traits because that "design" was self-evidently superior to all other possible "designs." The few species that separated from the main human lineage quickly went extinct, confirming the superiority of the main line. The idea that *Homo sapiens* is the natural, inevitable endpoint of all evolution is often implicit in popular culture.

Popular culture's best-known representation of human evolution reflects this optimistic vision. It shows, from the side, a line of figures marching across the page. At the far left a small, long-limbed, apelike creature seems to totter in its unfamiliar upright posture. At the far right a bearded man with a stone-tipped spear strides confidently forward. "Reading" from left to right, each of the figures is taller, more upright, and more human looking than the one before. The figures—often labeled with the names of their species—clearly represent successive stages in human evolution. Each is a descendent of, and an evolutionary step beyond, its predecessor. The progressive implications are underlined by the depiction of the figures walking rather than standing—like members of a parade rather than of a police lineup.

The widespread belief that humans evolved "from the apes" inevitably raises a troubling question: Just how different *are* we from the apes? Jocular answers have been part of popular culture since Darwin's *Origin of Species* entered bookshops in 1859. As early as 1861, cartoons in the British weekly *Punch* showed gorillas in evening dress attending high-society parties, and gorillas with protest signs demanding their civil rights. A widely repeated joke of the time has an image-conscious woman asking her husband about "Mr. Darwin's theory" that they are descended from apes. "Let us pray that it is not true," replies the husband, "and if it is true, let us pray that it does not become widely known!" Humor in a similar vein has flourished ever since. One edition of Gary Larson's comic strip *The Far Side* shows a caveman listening anxiously as a prospective employer expresses dismay that all the references listed on his resume are apes.

A much darker conclusion, that some of us are much more apelike than others, has also persisted from Darwin's day to ours. To be apish is to be less evolved, and so less human, than the average person. Nineteenth-century British cartoonists routinely drew Irishmen with apelike features, and their American counterparts did the same with blacks. During World War II, both Japanese and American propaganda artists drew apelike enemies. Applied to an individual, "ape" is standard American slang for a big, brainless lug. Bugs Bunny, in his distinctive Brooklyn accent, often addresses his muscle-bound enemies as "Ya big ape," and the sailors who do traditional seamen's jobs on American warships are "deck apes" to shipmates with more high-tech assignments. The title character of Eugene O'Neill's play *The Hairy Ape* is a burly stoker from

"The Darwin Club," drawn by Rea Irvin for the 18 March 1915 issue of *Life*. It was part of a series titled "Clubs We Do Not Care to Join." Courtesy of the Library of Congress.

the engine room of an ocean liner who becomes obsessed with a beautiful woman from First Class.

The fact that humans (even supposedly apish ones) are *not* apes raises another issue. The gap between humans and apes cannot have been crossed in a single evolutionary step; what lay between them? The standard answer revolves around a mythical "missing link." More than ape but initially less than human, it is usually pictured as a hairy creature with a slender humanlike body and a semi-erect posture. It "speaks" in grunts and apelike gestures, but a spark of humanness glows behind its eyes. Stories about the missing link tend to focus on a pivotal event that fans the spark into a flame and sets the missing link decisively on the evolutionary path that leads to us.

The pivotal event is always a visible, tangible action—never a slow, imperceptible process like the slow accumulation to genetic traits that actually drives evolution. Sometimes the missing link's role in the event is passive, as in modern legends about aliens who manipulated the course of evolution in order to produce us. Far more often, however, it is the missing link who takes action: he stands tentatively upright, he speaks his first halting "words," he hacks at an animal carcass with a sharpened stone. The "Dawn of Man" sequence that opens Stanley Kubrick's film *2001: A Space Odyssey* (1968) touches both bases. Subconsciously guided by a mysterious alien artifact, a tribe of apish creatures discovers that they can use animal bones as both tools and weapons. Their leader, giddily triumphant, throws a bleached leg bone high into the air. The camera follows it upward, then cuts to a winged spacecraft entering Earth orbit a million years later. Making tools, Kubrick suggests, made us human.

Ultimately the difference between these views of human evolution lies in their different perspectives on humans. The general public sees humans as uniquely well-designed creatures—the pinnacle of evolution. Scientists see them as one among thousands of successful species—as well adapted to their environments as the cactus, cod, and cockroach are to theirs. Placing *Homo sapiens* at the peak of an imagined evolutionary ladder says more about human vanity than it does about human biology.

Related Entries: Evolution; Prehistoric Humans; Prehistoric Time

FURTHER READING

Diamond, Jared. *The Third Chimpanzee*. HarperCollins, 1992. A series of loosely connected essays on human evolution, including language, behavior, and culture as well as anatomy.

Landau, Misia. *Narratives of Human Evolution*. Yale University Press, 1991. Argues that theories of human evolution mirror age-old folktales.

Tattersall, Ian. *The Fossil Trail: How We Know What We Think We Know about Human Evolution*. Oxford University Press, 1995. A concise, accessible survey of tools, bones, and DNA analysis.

Experiments

The close observation of nature has been part of Western science since its beginnings in the fifth century B.C. Experiments—observations made under controlled conditions—are much newer, products of the Scientific Revolution of the seventeenth century. They quickly became integral to science, however, and remain so to this day; every major scientific discipline makes at least some use of them. Experiments are also integral to science education. Beginning students carry out programmed experiments with known results—learning, in the process, that they can learn about nature by manipulating it. Advanced students are expected to design and carry out experiments of their own—demonstrating thereby that they have mastered an essential scientific skill.

Experiments typically take place in an environment—test tube, petri dish, cage, or vacuum chamber—physically separate from the rest of the world. Scientists can alter conditions within that environment to suit their needs. They can, if the space is suitably equipped, alter its temperature, its light levels, and the pressure and composition of its atmosphere. They can regulate what enters and leaves it. They can allow processes within its boundaries to unfold at will or, if they desire, introduce new elements that may alter those processes. Even the most basic experiments used in high school science classes—refracting light with a prism, rolling balls down variously angled inclines, identifying an unknown white powder—depend on controlling and manipulating a small section of the natural world.

This comprehensive control streamlines the process of observation and gives experiments much of their value. Carefully designed experiments allow scientists to observe efficiently the behavior of an object, organism, or process under a wide range of conditions. They allow scientists to create, on a small scale, conditions that are rare or nonexistent in nature or that occur only in inaccessible places. Experiments also allow scientists to test efficiently new theories. A well-crafted theory not only accounts for existing observations but allows scientists to predict behavior not yet observed. If a theory predicts that A will lead to B under conditions X, Y, and Z, experiments that place A under conditions X, Y, and Z offer a convenient way to compare it to observed reality.

The seventeenth-century founders of experimental science saw it as a way of gaining knowledge that would allow humans to manipulate nature for their own benefit. It gave scientists access to secrets that might otherwise lie hidden and allowed them to discover efficiently what kinds of manipulation produced the most valuable results. Seventeenth-century scientists reveled in both forms of control, but modern popular culture takes a more complex view. It distinguishes, fervently but not always clearly, between "good" experiments that serve all of humankind and "evil" experiments that serve a selfish few.

"Good" experiments typically produce not just knowledge but some exotic new substance. In fact, they often produce the substance by accident, leaving the experimenter ignorant of what it is or how it came to be. They are similar in that respect to the hallowed act of inventing a new machine, but they take place in a very different context. The person responsible is not a colorfully eccentric inventor or a mechanic tinkering in his spare time but a bookish, intense, serious scientist in a white lab coat. The setting is not a factory floor or basement workshop but an austere and forbidding laboratory. The tools involved are not familiar hammers, saws, and wrenches but glassware bent into exotic shapes and filled with strange substances. Even good experiments, then, take place in an atmosphere that suggests wizards at work and keeps ordinary mortals at arm's length.

The products of "good" experiments, however, have uses that are readily comprehensible to ordinary folk. They provide detectives from Arthur Conan Doyle's Sherlock Holmes to Patricia Cornwell's Dr. Kay Scarpetta with the insights needed to solve baffling crimes. They provide the means to combat deadly diseases, as in Sinclair Lewis's *Arrowsmith* (novel 1922, film 1931) and Michael Crichton's *The Andromeda Strain* (novel 1969, film 1971). They produce (notably in movies) objects with undreamed-of properties: wood-avoiding baseballs in *It Happens Every Spring* (1949), stain-resistant fabric in *The Man in the White Suit* (1957), and flying rubber in *The Absent-Minded Professor* (1962). They yield (again, primarily in movies) potions that transform the experimenters, conferring qualities such as youth (*Monkey Business*, 1952), invisibility (*Now You See Him, Now You Don't*, 1972), or strength (*The Strongest Man in the World*, 1975).

"Bad" experiments are the work of experimenters who have lost their moral bearings and—driven by greed or ambition—turned their attention to parts of nature that humans have no business manipulating. Bad experiments are almost invariably biological, and (like good experiments) they almost invariably produce tangible results. Such experiments are bad, popular culture typically suggests, because the creation or radical modification of life is the province of God or nature, not of fallible humans. Mary Shelley's novel *Frankenstein* (1818) established the basic plot and tone used in most bad-experiment stories of the two centuries. In it, an overly ambitious scientist creates a new life form that, he discovers too late, neither behaves according to his expectations nor submits to his control. The creation grows vengeful and threatening, but in the end it is vanquished, and order is restored.

Victor Frankenstein's successors—"playing in God's domain" with no thought of the consequences—are stock characters in popular culture. The title character of H.G. Wells's novel *The Island of Dr. Moreau* (1896) tries to bridge the gulf between humans and animals. Scientists in the movie *Species* (1995) combine human and alien DNA. Governments and corporations use unsuspecting citizens as experimental guinea pigs in Stephen King's novel *Firestarter* (1980) and TV series such as *Dark Angel* (2000–). The experiments depicted in Daniel Keyes's "Flowers for Algernon" (1959, expanded 1966) and Michael Crichton's *The Terminal Man* (novel 1972, film 1974) have noble goals but end tragically because of the scientists' arrogance. Depictions of bad experiments that depart from the *Frankenstein* model are comparatively rare. Michael Crichton's novel *Jurassic Park* (1990) was one such rarity: its central message is the danger of allowing experiments with genetic engineering to take place, unobserved and unregulated, behind the closed doors of corporate labs. Stephen Spielberg's 1992 film moves the story back into familiar *Frankenstein* territory, suggesting that genetic engineering upsets the natural order of things and is therefore wrong by definition.

The experiments depicted in popular culture, whether "good" or "bad," typically take place off stage or in the background. This is true even in realistic accounts of actual experiments, such as those in "reality TV" shows focused on medicine and forensic science. The results of the experiment take pride of place, appearing in the foreground with clear voice-over explanations of what they mean. Failed experiments, ambiguous results, and the tedious but vital work of preparing specimens, taking measurements, and writing up results are seldom part of the story. Popular culture thus creates a cumulative image of experimentation as a quick, efficient, virtually infallible way of uncovering nature's secrets and using them to solve problems. It is a seductive image, but one in which few scientists recognize day-to-day realities of their work.

Related Entries: Experiments on Self; Scientific Theories

FURTHER READING AND SOURCES CONSULTED

Collins, Harry M., and Trevor Pinch. *The Golem: What Everyone Should Know about Science*. Cambridge University Press, 1993. Seven case studies of the interplay of theory and experiment.

Gould, Stephen Jay. "Jurassic Park." In Mark C. Carnes, ed., *Past Imperfect: His-*

torians Look at the Movies. Henry Holt, 1995. Analyzes the differences, mentioned here, between novel and film.

Haynes, Roslynn D. *From Faust to Strangelove: Representations of the Scientist in Western Literature*. Johns Hopkins University Press, 1994. The definitive treatment of fictional experimenters.

Merchant, Carolyn. *The Death of Nature: Women, Ecology, and the Scientific Revolution*. HarperCollins, 1980. A controversial work on the origins of modern science; source of the experiment-as-control metaphor used throughout this entry.

Experiments on Self

Regulations covering scientific experiments on human subjects—shaped by the memory of past excesses—are stringent. Subjects must give their informed, written consent, and experimenters must submit their proposed experiments to rigorous scrutiny by overseeing bodies. Scientists who experiment on themselves can, functionally if not legally, avoid the restrictions associated with experimenting on other people. They can also sidestep most of the ethical issues involved: nobody, presumably, is more aware of an experiment's potential hazards than the scientist who devised it.

Nonetheless, experimenting on oneself remains deeply problematic. One obvious drawback is the danger involved; knowing that it exists does nothing to reduce it. A less obvious drawback is the limited range of data that the experiment can generate. Human anatomy and physiology vary, in small but significant ways, according to sex, age, lifestyle, and other factors. Experimental results derived from a single subject are, therefore, of limited value; there is no way to know whether the subject's responses are typical or atypical of the response of humans as a group. Finally, scientists who experiment on themselves may suffer physical or psychological effects that compromise their ability to record and analyze data objectively.

Auto-experimentation (its formal name) carries, despite these drawbacks, seductive overtones of edginess and daring. Its most famous real-world practitioners enjoy, among those who know their work, a folk-hero status that few of their colleagues can claim. Air force physician John Stapp is famous for strapping himself to rocket-driven sleds in order to measure the human body's ability to withstand rapid acceleration and deceleration J.B.S. Haldane, one of the leading evolutionary biologists of the mid-twentieth century, saw auto-experimentation as both scientifically valuable and (implicitly) as a test of personal courage. He compared the auto-experimenter to "a good solider who will risk his life and endure wounds in order to gain victory," and he saw death by auto-experiment as "the ideal way of dying" (Oreskes 107).

Depictions of auto-experimentation in popular culture emphasize its dangers rather than its perceived glamour. Fictional scientists who experiment on themselves often suffer death, permanent disfigurement, or disturbing personality changes. Their experiments are scientific failures that teach them more about their own limitations than about the secrets of the universe. The experiment thus reinforces, in a deeply personal way, a standard theme in popular culture's treatment of science: that manipulating nature is dangerous.

Tales of auto-experiments gone wrong are usually horror stories in which the hero loses control of his body, his mind, or both. The side-effects of an experimental potion turn a quiet physician into a brutal murderer in Robert Louis Stevenson's *The Strange Case of Dr. Jekyll and Mr. Hyde* (1886). Experimenting with invisibility brings out the worst in H.G. Wells's *The Invisible Man* (novel 1897, film 1933), turning him too into a murderer. The psychologist-hero of the surrealistic movie *Altered States* (1981) suffers "evolutionary regression" under the influence of hallucinogenic drugs and an isolation chamber. He becomes, first in mind and then in body, one of humankind's distant hominid ancestors. Both movie versions of *The Fly* (1958, 1986) use a botched auto-experiment to scramble the hero's body parts with those of the titular insect. The *Man with X-Ray Eyes* (1963) greets the godlike powers his experiment gives him first with delight and later with horror. Finally, desperate, he seeks relief by tearing out his own eyes.

Stories of auto-experimentation are not uniformly grim or horrifying. Even those intended as comedies, however, place the hapless experimenters in (comic) danger or subject them to public embarrassment. The plots of the comic stories are, in fact, often thinly disguised retreads of *Dr. Jekyll and Mr. Hyde* or *The Invisible Man*, with romantic complications substituted for violence and humiliation for disfigurement. The 1996 remake of *The Nutty Professor*, for example, reaches its climax when Professor Sherman Klump (Eddie Murphy) reverts—in a crowded room—from the superstud alter-ego created by his "secret formula" to his sweet, bumbling, overweight "real self."

Only in superhero adventures do fictional auto-experimenters display the heroism attributed to their real-world counterparts. Dr. Peyton Westlake fights crime as "Darkman" in a series of movies (1990, 1995, 1996), hiding his disfigured face behind artificial skin that he invented. Marvel Comics' "Ironman" and the hero of the TV series *M.A.N.T.I.S.* (1994–1995) enclose their crippled bodies in elaborate mechanical shells of their own design. These heroes, however, are auto-experimenters only in the broadest sense of the word. The mechanical devices they apply to their bodies augment them without penetrating or permanently modifying them. Their experiments carry no risk of death or permanent injury. Indeed, they are designed to compensate for the effects of existing injuries. All three turn to auto-experimentation only after their bodies are already badly damaged—only when extreme and desperate measures are required.

Related Entries: Experiments; Mutations

FURTHER READING

Altman, Lawrence K. *Who Goes First? The Story of Self-Experimentation in Medicine.* Random House, 1987. A comprehensive history of auto-experimentation in medicine.

Clark, Ronald W. *JBS: The Life and Work of J.B.S. Haldane.* Hodder and Stoughton, 1968. A biography of the twentieth century's best-known auto-experimenter.

Oreskes, Naomi. "Objectivity or Heroism? On the Invisibility of Women in Science." *Osiris*, 2nd series, 11 (1996): 87–113. Discusses heroism in science, with auto-experimentation as one theme.

Extinction

Scientists and the general public define extinction the same way: as the "death" of an entire species. The two groups, however, think about extinction in very different contexts and understand it in very different terms.

Scientists see extinction as a natural result of the complex interactions between living creatures and their continually changing physical environment. When a species becomes so ill suited to its environment that it cannot compete successfully for food and living space, it is likely to become extinct. The crisis may be brought on by evolutionary changes in the species itself, by changes in the environment, or by the arrival of a new species (predator or efficient competitor) in the same local area. Large-scale environmental changes can cause "mass extinctions" that wipe out enormous numbers of species in a geologically short time. The extinction of a single species is, from a scientific standpoint, like a single dot of color in an impressionist painting—significant primarily as a part of a larger pattern.

Popular culture, on the other hand, invests the extinction of individual species with deep meaning and its stories about extinctions with a strong moralizing tone. Those set in the present portray threatened species as sweet, innocent victims of human ignorance and greed. Those set in the past portray now-extinct species as hapless victims of their own shortcomings. Both portrayals are rooted in broader cultural attitudes toward nature.

Stories about impending extinctions, whether told as entertainment or as calls to political action, invariably side with the threatened species. They almost invariably focus on what one biologist has called "charismatic megafauna": big, familiar, photogenic mammals like tigers, gorillas, and whales. Such species are, culturally if not biologically, more "human" than rodents, fish, and frogs. They are thus easier to generate human sympathy for. Fiction about threatened species—from children's

Name	Years Ago	Species Killed
Precambrian	650 million	70%
Cambrian	510 million	60%
Ordovician	440 millin	75%
Devonian	368 million	70%
Permian	248 million	95%
Cretaceous	65 million	85%

Major mass extinctions in Earth history. The "Species Killed" column lists the percentage of *all* species then living that became extinct. Extinction rates for particular groups of species may be higher. The Cretaceous mass extinction, for example, killed every terrestrial species weighing more than fifty-five pounds. Data from Raup and "Extinction."

books like Dr. Seuss's *The Lorax* (1971) to adult novels like Hank Searls's *Sounding* (1982)—routinely enhances the animals' appeal by giving them humanlike personalities and the power of speech. Even in nonfiction—notably direct-mail appeals from environmental organizations—the animals are routinely shown in tight close-ups, looking into the camera like a person sitting for a portrait. The stories, whether fiction or nonfiction, recall countless tales of anthropomorphic animals. A call to protect dolphins is, at some level, a call to protect Flipper. An appeal to save endangered tigers calls up, however irrationally, memories of Kellogg's Frosted Flake's Tony and Winnie-the-Pooh's bouncy friend Tigger.

Popular culture's treatment of past extinctions takes precisely the opposite view, arguing that the now-extinct species deserved their fate. This attitude is rooted in a linear, progressive view of evolution that was common among nineteenth-century scientists and remains common in popular culture. Evolution, according to this view, is a steady climb along a more or less predetermined "ladder" from simplicity and crudeness to complexity and sophistication. An extinct species is one that has failed to evolve, failed to evolve quickly enough, or evolved along a "dead-end" path leading away from the "main line." It cannot compete with more advanced species that kept moving briskly along the evolutionary main line, and so disappears. The linear-progressive view implies that evolution is driven by forces (never well defined) within each species. Failure to evolve properly is thus a "personal" failure on the part of the species, a lack of will or ambition. Extinction is, by this definition, *always* the victim's fault.

Popular culture abounds with examples of this attitude at work, examples that often contrast evolution's extinct "failures" with successful humans (or their ancestors). The dodo bird's eradication by human hunters in the seventeenth century is, in popular culture, proof of its evolutionary "failure." Human hunters have slaughtered other species in far greater numbers, but the dodo, by becoming extinct, became an emblem of fatal stupidity. Dinosaurs, the best known of all extinct animals, receive similar treatment in popular culture. They appear, juxtaposed with the small mammals that were their contemporaries and successors, in narratives similar to folktales about big, dumb villains beaten by small, clever heroes. The agile, intelligent dinosaurs featured in the film *Jurassic Park* (1992) have done little to change this tradition. "Dinosaur," when used metaphorically to label people or organizations, is never complimentary. It connotes backwardness, sluggishness, and lumbering stupidity—a creature too big and clumsy to survive and too stolid to adapt. The dinosaurs' demise, like the mammals' triumph, is depicted as an inevitable, necessary step on the road to a more biologically advanced world.

Related Entries: Dinosaurs; Evolution

FURTHER READING AND SOURCES CONSULTED

"Extinction: Cycles of Life and Death through Time." 19 December 1996. Hooper Virtual Paleontology Museum. 23 August 2001. <http:// hannover.park. org/Canada/Museum/extinction/tablecont.html>. Bite-sized introductions to the principal extinctions in Earth history.

Hsu, Kenneth J. *The Great Dying*. Harcourt Brace Jovanovich, 1986. Uses the Cretaceous-Tertiary extinction that destroyed the dinosaurs to illustrate larger themes about the role of extinction in evolution.

Raup, David. *Extinction: Bad Genes or Bad Luck?* Norton, 1992. Useful overview of why species become extinct, by an expert in the field.

Ward, Peter Douglas. *Rivers in Time: The Search for Clues to Earth's Mass Extinctions*. Rev. and updated ed. Columbia University Press, 2001. A comprehensive general introduction.

Flying Cars

The image of personal air transportation is one of irresistible simplicity. Step into your garage, climb into your vehicle, and in moments you can be zipping through the skies on a journey free of potholes, toll booths, traffic jams, and stoplights. Every journey could be by the most direct route possible, literally "as the crow flies." Remote areas without roads—or with roads too steep, rough, or winding for ordinary cars—would be as accessible as the heart of downtown. This image has been part of popular culture virtually since the invention of flight in the early twentieth century. Flying cars—actually, coupes with detachable wings and tails that turned them into airplanes—enjoyed limited popularity and extensive press coverage in the United States between 1945 and 1960. Designers are currently pursuing a half-dozen major flying-car projects, of which Paul Moller's M400 Skycar (computer-controlled and capable of vertical takeoffs) is the best known. The technology of flying cars has changed, but the dream underlying them has not.

The realities of creating personal, everyday air transportation—"flying cars," for short—are in fact nightmarishly complex. A heavier-than-air flying machine must have a light but durable structure and a light but powerful engine in order to work at all. The engine must also be reliable, since its failure in flight is by definition a life-threatening crisis. A flying machine designed for daily use by an operator with limited training and experience would also have to be extremely stable, both in level flight and during maneuvers. Solving these technological problems would set the stage for putting "a flying machine in every garage." That, in turn, would create new social problems: air traffic control; three-dimensional "rules of the road"; insurance; training, licensing, and inspection; and liability for damage done on the ground by machines that crash.

Popular culture focuses on flying cars' glittering image and carefully side-steps their complex realities. It treats them as the highest of high technology, symbols of technology's power to improve everyday life.

Unlike other symbols of the high-tech future—robots, smart houses, and intelligent computers—flying cars are never shown breaking down, much less developing malevolent agendas of their own. They are as reliable and easy to use as a telephone or refrigerator is in our world and, in stories set in the future, as familiar and unremarkable to their users.

Personal flying machines figured prominently between 1910 and 1940 in future cities like those in *Metropolis* (film 1926) and *Things To Come* (film 1936). Artists' conceptions of such cities show a wide range of craft—airplanes, dirigibles, and the occasional helicopter—moving in orderly streams above multitiered highways and between towering skyscrapers. Their typical occupants, elegantly dressed as though for an evening at the theater or a fine restaurant, are clearly using their machines for routine in-town transportation. Flying cars are even more prevalent in *The Jetsons*, a 1962 animated TV series about a "typical family" of the twenty-first century. Bubble-topped machines, apparently jet powered and computer controlled, have completely replaced the family station wagon (and, apparently, all other forms of ground transportation). The Jetson family's car is capable of supersonic speeds and rock-steady hovering, and it folds itself into a briefcase when not in use. Typically, for stories about the future, it is ordinary to its users and astonishing to twentieth-century audiences. *Blade Runner* (1982) is nominally set in the first decade of the twenty-first century, but its nightmarish vision of a neon-lit, pollution-choked, perpetually rainy Los Angeles seems to belong to a more distant future. Residents are unfazed, therefore, when flying cars swoop overhead and a flying police cruiser descends vertically into their midst.

Popular culture's association of flying cars with the future is strong even in stories that take place in the present. Flying cars are, in such stories, treated as machines so advanced they seem almost magical. Characters who build their own flying cars—like the eccentric inventors in *The Absent-Minded Professor* (1962), *Chitty Chitty Bang Bang* (1968), and the *Back to the Future* films (1985–1990)—are treated as modern-day wizards. Characters who merely *use* flying cars are invariably shown to be partners of a wizardlike inventor or beneficiaries of some other form of advanced intelligence. The jet backpack that James Bond uses at the beginning of *You Only Live Twice* (1967) is the brainchild of Q, the Secret Service's endlessly inventive armorer. The flying bicycles in the climax of *E.T.* (1982) fly because the wide-eyed alien of the title wills them to.

The idea that technology will magically transform us and our lives has been part of Western culture since the beginnings of industrialization. Flying cars, magically sweeping away the drudgery and frustration of driving, were enduring symbols of that idea in the twentieth century and remain so in the twenty-first.

Related Entries: Computers; Food Pills; Houses, Smart; Intelligence, Artificial; Robots

FURTHER READING AND SOURCES CONSULTED

Corn, Joseph J. *The Winged Gospel: America's Romance with Aviation, 1900–1950.* Oxford University Press, 1983. Chapter 5, "An Airplane in Every Garage?" covers that persistent dream.

Grossman, John. "Auto Pilots." *Air and Space/Smithsonian*, January 1996, 68–77. Comprehensive survey of flying car designs from their brief post-1945 heyday to their current semirevival.

White, Randy Wayne. "Would You Buy a Flying Car from This Man?" *Men's Health* 15: 4 (May 2000): 66–71. Profile of Paul Moller, designer and promoter of the Moller Skycar and a passionate advocate of mass-marketed flying cars.

Food Pills

We eat for nourishment, but not only for nourishment. Eating can be a form of sensual self-indulgence (as countless odes to chocolate make clear) or a form of self-medication (as any habitual coffee-drinker can attest). Eating is central to religious rituals such as the Jewish seder and Christian Eucharist. Meals taken together strengthen the social bonds within communities, families, and couples. Many foods carry strong cultural connotations about the race, class, and gender of those who relish them: Royal Crown cola and Moon Pie pastries have been called, only half-jokingly, the champagne and caviar of the southern white working class. Specific foods also remain, despite the rise of chain restaurants with standardized menus, central to the self-image of particular regions, states, and even cities. New England has its milk-based clam chowder, Maryland its crab cakes, Chicago its deep-dish pizza. Finally, certain foods carry emotional connotations that transcend individual diners' memories of them. Ending an intimate dinner for two with a rich chocolate dessert sets a different mood than ending it with, say, sliced watermelon.

Compact food—explorers' dehydrated meals, soldiers' field rations, athletes' energy bars, dieters' meal-replacement drinks—provides maximum nourishment with minimum weight and bulk. Beyond that sometimes-vital combination, it offers little. It tends, in all its forms, to lack the taste, texture, color, and variety of conventional food, and very few who eat it regularly prefer it to the "real thing." American soldiers joked in the 1990s that the "MRE" label on their ration packs stood not for "Meals, Ready-to-Eat" but for "Meals Rejected by Everyone" or "Meals Rejected by the Enemy." Even in the field, they are traditionally laid aside on holidays in favor of meals prepared with "real food" and traditional recipes. The resulting morale boost is twofold: the holiday meals taste better, and they are a powerful symbol of home.

Food pills, imagined but not yet developed, would be the ultimate form of compact food: an entire meal that could be consumed in a single swallow. Variations in color, taste, texture, and smell would presumably be irrelevant in a food pill, except as a marketing device. Variations in content—to tailor the pills to the age, sex, lifestyle, and medical condition of the consumer—might well be indicated (as they are now in vitamin supplements) only on the packaging. Food pills would offer those who ate them neither sensual pleasure, an opportunity to socialize, nor a chance to make a symbolic statement. They would reduce the act of eating, once complex and multidimensional, to its most basic purpose: nourishment.

Food pills have a dual presence in popular culture. The pills often appear in science-fictional settings where good nutrition must be maintained but weight and bulk must be strictly limited. The astronauts in the 1955 film *The Conquest of Space* dine on food pills because, presumably, more conventional food would be too bulky to carry on their long journey to Mars. Major Tom, the ill-fated hero of David Bowie's 1969 song "Space Oddity," swallows protein pills as he prepares to enter orbit in his cramped one-man spacecraft. The plot of the film *Soylent Green* (1973) revolves around a protein wafer used to feed the masses efficiently in a drastically overpopulated New York City of the near future. Food pills function in these stories the way compact food does in the real world: as a substitute for "real food," used only when circumstances require it.

Food pills also appear, however, as the staple food of the future, consumed by everyone regardless of wealth or occupation. The pills—along with flying cars, automated houses, and robot servants—became part of a standardized image of the future outlined in magazine articles during the late 1940s and 1950s. These portrayals of the future never explained, or even implied, *why* typical American families would forego the pleasures of waffles, hot dogs, and apple pie in favor of pills. Little else in the portrayals hinted at the hurried lifestyles that would make two-minute meals attractive. These contradictions hastened the disappearance of food pills from serious attempts to imagine the future.

The memory of food pills lasted far longer—sustained, paradoxically, by its perceived absurdity. The TV cartoon series *The Jetsons* (1962–1963) used them in its gentle mocking of the 1950s' imagined future, and Woody Allen's film *Sleeper* (1973) used them to satirize the contemporary health-food craze. End-of-century commentators regularly seized on them as an example of how widely the 1950s' vision of the future had missed the mark. This shift in attitude reflects a basic truth about food pills: that the problems of compressing a balanced "meal" into a capsule are dwarfed by the problems of divorcing eating as nourishment from the many other aspects of eating.

Related Entries: Flying Cars; Houses, Smart; Robots

FURTHER READING

Beardsworth, Alan, and Keil, Theresa. *Sociology on the Menu: An Invitation to the Study of Food and Society.* Routledge, 1997. An accessible introduction to the social dimensions of eating.

Corn, Joseph J., Brian Horrigan, and Katherine Chambers. *Yesterday's Tomorrows: Past Visions of the American Future.* Johns Hopkins University Press, 1996. A well-illustrated history of ideas about the future, including food pills.

Pillsbury, Richard. *No Foreign Food: The American Diet in Time and Place.* Westview Press, 1998. A sweeping overview of what Americans eat and why they eat it.

Franklin, Benjamin

Science was not, in the mid-eighteenth century, professionally or intellectually separate from philosophy and other learned activities. Extensive professional training, specialized university degrees, and expectations of paid employment were not yet standard elements of scientific careers. Many of the leading "scientists" of the era combined their research with full-time jobs in other fields. These limitations did not, however, prevent them from making significant discoveries and developing important ideas. Benjamin Franklin (1706–1790) fit both patterns. Though not a career scientist, he did substantial work on one of the leading problems in eighteenth-century physics, the nature of electricity.

Franklin's first important contribution to the science of electricity was his discovery that a slender rod made of conductive material could, if held close to a charged object, draw off the charge without touching the object. This discovery led Franklin to perform his famous 1752 kite experiment, which showed that lightning was a naturally occurring form of electricity. It also led him to develop the lightning rod, which he believed, over-optimistically, could "defuse" thunderclouds by drawing off their electrical charge before it could become lightning. The kite experiment, though spectacular, was not original. It had been suggested, and performed, in Europe as early as 1749.

Franklin's second major piece of electrical work was more original. Most eighteenth-century scientists envisioned electricity as a weightless fluid created by rubbing certain materials. It came, they believed, in two forms: "vitreous" (created by rubbing glass) and "resinous" (created by rubbing amber). Franklin, on the basis of a series of experiments, proposed a different theory. Electricity, he argued, was a single fluid that could be neither created nor destroyed. Friction could, under the right circumstances, transfer it from one part of an object to another, creating a surplus in one place and a deficit in another. Franklin dubbed the two kinds of charge "positive" and "negative"—terminology still in use to-

Franklin's Kite Experiment, painted by Charles E. Mills around 1911. The bottle-shaped device to his left is a Leyden jar, an eighteenth-century device for storing electrical charge. Courtesy of the Library of Congress.

day—and used his "single fluid" theory to explain most of the properties of electricity then known. His belief that electricity was conserved (neither created nor destroyed) was one thread that led to the generalized conservation-of-energy law developed in the 1840s.

Popular culture dotes on Franklin but pays little attention to his science. Only the kite experiment receives more than passing notice, notably in brief segments of Walt Disney's "Ben and Me" (1953) and Warner Brothers' "Yankee Doodle Bugs" (1954). Popular culture's depictions of the kite experiment are brief and context free, more spectacle than science. They present it as an isolated "bright idea" devised and executed by Franklin alone, not as part of a research program that connected Franklin to fellow scientists both in America and in Europe, and they present it in heavily edited form. They omit the silk ribbon that insulated Franklin from the key, reducing death from a certainty to merely a possibility. They often fail to credit Franklin and his contemporaries with even knowing about electricity beforehand. Hence, after the kite and key have done their work, Franklin is often said to have "discovered electricity."

The narrow depiction of Franklin's science in popular culture is, in part, a practical matter. An experiment with ordinary objects is easier to depict on screen or explain on paper than one with now-unfamiliar laboratory apparatus like Leyden jars. Almost *any* experiment, moreover, is easier to depict or explain than an abstract theory, even a relatively familiar one like Franklin's single-fluid model of electricity. Popular culture's limited attention to Franklin's science also has other roots, however.

Franklin's life and personality differ, in every way imaginable, from popular culture's standard image of a scientist. Scientists (according to popular culture) are cold and unemotional; Franklin was jovial and pleasure loving. Scientists have difficulty relating to other people; Franklin made a career of doing so. Scientists are abstract, distant, and impractical; Franklin founded schools and fire departments, invented bifocals, and improved the cast-iron stove. Scientists have no interest in the world outside the laboratory; Franklin helped to found a nation. It is impossible to imagine moon-faced, avuncular Howard da Silva (who played Franklin on stage and screen in *1776*) cast as a scientist in *Frankenstein* or *Jurassic Park*.

Popular culture's depiction of the kite experiment reinforces its image of Franklin as an American icon, not a scientist. It presents Franklin's "science" as something very different from what most scientists in popular culture do. Franklin's science is experimental, not theoretical. He does it with familiar objects, not strange-looking laboratory equipment. Most important, he does it at great personal risk: alone, outdoors, at the height of a thunderstorm. The kite experiment thus reinforces Franklin's traditional status as a model of distinctly American values: practicality, resourcefulness, and personal courage. It establishes him as a True American by distancing him from the theory-laden, laboratory-bound, reality-shy world of the scientist.

Related Entries: Electricity; Lightning

FURTHER READING AND SOURCES CONSULTED

Brands, H.W. *The First American: The Life and Times of Benjamin Franklin.* Doubleday, 2000. A comprehensive, readable biography that places Franklin's science in the context of his own life and other interests.

Cohen, I. Bernard. *Benjamin Franklin's Science.* Harvard University Press, 1996. Collected essays by the foremost scholar of Franklin's science, arguing for his role as theorist, as well as an inventor.

Hankins, Thomas L. *Science and the Enlightenment.* Cambridge University Press, 1985. A useful brief (pp. 62–67) discussion of Franklin's electrical work in the context of eighteenth-century experimental physics; source of the historical material in this entry.

Galileo

Galileo Galilei (1564–1642) was one of the leading figures in the 'Scientific Revolution" that, between roughly 1550 and 1700, transformed educated Europeans' understanding of the natural world. The centerpiece of the revolution was the formulation of a new system of physics and astronomy to replace those introduced by Aristotle 2,000 years before. Galileo did important work in both astronomy and physics, but popular culture—and so this entry—focuses on the former.

The Aristotelian worldview placed the earth at the center of the universe, orbited by the moon, sun, and planets. The orbit of the moon separated the celestial realm of changeless perfection from the terrestrial realm of change and decay. It placed humans at the center of the universe, but in a world the imperfection of which reflected their sinful nature. Galileo, on the other hand, erased the terrestrial-celestial distinction and placed the earth among the planets, orbiting a fixed sun. The model was devised by Nicholas Copernicus in 1543, but Galileo—whose career began a half-century after Copernicus died—became its most zealous promoter (Alioto 64–68, 178–185).

One crucial element of Galileo's campaign was the telescope, a new instrument that he was the first to use for astronomy. Some of what he saw—Earthlike mountains on the moon and spots on the sun—eroded the Aristotelian idea of a "perfect" celestial realm. Other discoveries, such as the four largest moons of Jupiter, lent plausibility to the idea that the moon could orbit a moving Earth. Galileo's discovery that Venus, like the moon, had phases was crucial. It simply could not happen if, as in the Aristotelian model, Venus orbited the earth. Galileo's observations supported, but did not prove, the Copernican model. All of them could be accounted for equally well by a third model, proposed by Danish astronomer Tycho Brahe (1546–1601), in which the moon and sun orbited a stationary Earth while the planets orbited the sun (Alioto 212–218). The Tychonic model *looked* far less elegant than Copernicus's, but

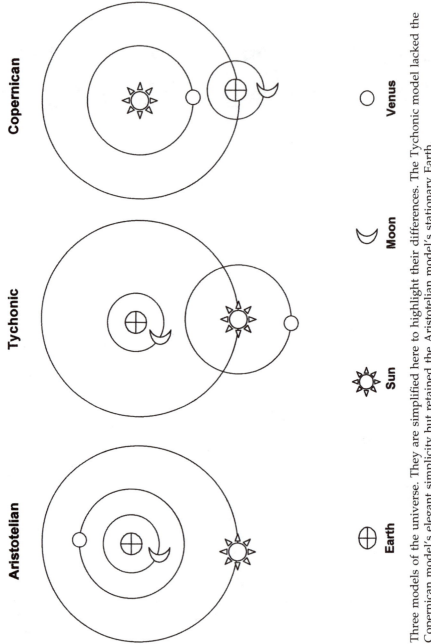

Aristotelian

Tychonic

Copernican

⊕ **Earth**

☀ **Sun**

☾ **Moon**

○ **Venus**

Three models of the universe. They are simplified here to highlight their differences. The Tychonic model lacked the Copernican model's elegant simplicity but retained the Aristotelian model's stationary Earth.

it tracked the motions of celestial bodies equally well. It also allowed conservative scholars to keep Earth stationary at the center of the universe: a position befitting the home of God's most important creation, and one that did not (as Copernicus's moving Earth did) render Aristotle's comprehensive system of physics obsolete (Alioto 184–188).

Galileo's campaign was also complicated by his zealousness. His attacks on scientific orthodoxy made powerful enemies, particularly among priests of the Jesuit order, who acted both as scholars and as defenders of the faith. When in 1632 he made the dangerous misstep of appearing to ridicule the pope in print, his enemies moved decisively against him. Brought to trial before the Inquisition and charged with holding and teaching heretical ideas, he was forced to recant publicly his belief in a sun-centered universe. His enemies thus achieved their goal. Galileo, a public figure proud of his image and reputation, had been publicly humiliated (Shea 124–133).

The Galileo of popular culture is less a person than a symbol. Occasionally, as in the revised, post-Hiroshima version of Bertholt Brecht's play *Galileo* (1945), he is a dark figure: a calculating opportunist who publicly renounces the truth when it is expedient to do so. Brecht makes him the forerunner of twentieth-century physicists who, by choosing to develop nuclear weapons, sacrificed their principles at the altar of political expedience and government funding. More often, as in Robert Heinlein's young-adult science fiction novel *Rocket Ship Galileo* (1947), he is a paragon of scientific virtue. After naming their spaceship, the three teenaged heroes remember its namesake as a man "whose very name has come to stand for steadfast insistence on scientific freedom and the freely inquiring mind."

Popular culture places the reader or viewer firmly on Galileo's side. Brecht, Heinlein, and virtually everyone else who invokes him do so with the benefit of hindsight. They take for granted that Copernicus was right—a position universally held in the twentieth century but not, even among scientists, in the seventeenth century. The omission of the Tychonic model from most popular accounts strengthens Galileo's position by making the Aristotelian model (clearly discredited by his telescope) the only alternative to the Copernican. Galileo thus appears in popular culture not in his historical role as the leading advocate for a controversial theory but in the idealized role of embattled truth teller. The nobility of Heinlein's Galileo lies in his embrace of that role (up to the last moments of his trial, which go unmentioned). The tragedy of Brecht's Galileo lies in his ultimate betrayal of it. One enduring—though probably apocryphal—story acknowledges Galileo's forced recantation before the Inquisition but gives him the last word by having him mutter "and yet [the earth] moves" as he leaves the courtroom.

Related Entries: Ideas, Resistance to; Newton, Isaac; Religion and Science; Scientific Theories

FURTHER READING AND SOURCES CONSULTED

Alioto, Anthony M. *A History of Western Science*. 2nd ed. Prentice-Hall, 1993. The best one-volume survey of the history of science; places Galileo's work in context.

Galilei, Galileo. *Discoveries and Opinions of Galileo*. Trans. and ed. Stillman Drake. Doubleday Anchor, 1957. A selection of Galileo's most important shorter works.

The Galileo Project. Ed. Albert Van Helden and Elizabeth Burr. 5 August 1996. Rice University. 12 November 2000. <http://es.rice.edu/ES/humsoc/Galileo/>. Information about Galileo's life, work, and world; with references.

Sharratt, Michael. *Galileo: Decisive Innovator*. Cambridge University Press, 1996. An excellent, concise biography designed for nonspecialists.

Shea, William R. "Galileo and the Church." *God and Nature*. Ed. David C. Lindberg and Ronald L. Numbers. University of California Press, 1986. Balanced treatment of the most complex episode in Galileo's life.

Genes

Genes are bits of material (mostly DNA) that encode and transmit specific hereditary information. They are usually, but not inevitably, located on the chromosomes, long, threadlike structures found in the nuclei of cells. Every species has a characteristic number of chromosomes (twenty-three pairs in humans) and a characteristic complement of genes (100,000 or so in humans), called its "genome." The genes that an organism inherits direct the formation of the proteins that make up its new cells as it grows and so define the physical form of the organism.

Working out that genes shape the physical form of organisms, and how, took scientists most of the twentieth century. Work done—or, in the case of Gregor Mendel, rediscovered—at the beginning of the new century established the basic principles. Midcentury investigations of the structure and function of DNA clarified the biochemical mechanism. The recently completed Human Genome Project mapped the location of each of humankind's 100,000 or so genes. We now know a great deal about human genes but very little about the roles played by particular genes in creating particular physical traits. Some traits have been correlated with particular genes, but the apparent success of that process (each new step of which is trumpeted in the media) is deceptive. Many apparent correlations have later turned out to be false—individuals without the trait were discovered to have the gene—and even a valid correlation does not establish causality. The gene in question may be necessary but not sufficient by itself to produce the trait. It may be one of a set of genes, common individually but uncommon in combination.

Understanding the genetic component of specific behaviors, as opposed to specific physical traits, is even more complex. Most scientists agree that behavior is the product of a complex interaction between genetics, physical environment, and social environment. Few agree specifically on the relative importance of the three factors, however, and fewer still claim to understand the details of how they combine to produce a specific result. The process is complicated further by the difficulty of defining specific behaviors. Arguing about the roots of intelligence means first agreeing on what "intelligence" *is*—a daunting task in itself. Many scientists have, since the mid-1970s, grown significantly more confident that behavior has a strong genetic component and that we will soon be able to discern the specific genetic roots of specific physical and behavioral traits. Richard Dawkins, Edward O. Wilson, and Matt Ridley (among others) have articulated these views with depth and subtlety in popular books. The same position has also come to dominate popular culture since the mid-1970s. Popular culture's version, however, typically offers neither depth nor subtlety. Driven by the need to tell exciting stories, it reduces genetics to two principles: that genes will prevail over any environmental influence, and that even complex traits have simple genetic roots. The result is a kind of genetic predestination. We are what our genes make us, popular culture suggests, and cannot escape their influence.

A genetic inclination to violence often served, in the last decades of the twentieth century, the same dramatic purpose that "fate" served in earlier centuries. Oedipus spends most of Sophocles' play *Oedipus Rex* (c. 430 B.C.) trying unsuccessfully to evade a prophecy that he will kill his father and marry his mother. The murderers in the controversial film *Natural Born Killers* (1994) are equally doomed but—as the title implies— by heredity rather than fate. Freddy Krueger, villain of the *Nightmare on Elm Street* series (1984–1994), was (according to events recounted in the first film and depicted in the fifth) born to a mother gang-raped in a mental institution. His penchant for gruesome murders is implicitly a product of his status as "the bastard son of a hundred maniacs." The movie *Alien*[3] (1992) makes its genetic determinism explicit; the prison planet where it is set is populated entirely by men whose possession of an extra *Y* chromosome (*XYY* instead of the traditional *XY* pair found in human males) predisposes them to violence. They cannot be reformed, since genetics is stronger than any environmental influence; they have been exiled instead.

Positive qualities, when genetically imprinted, are equally irresistible. *Star Wars, Episode I: The Phantom Menace* (1999) explains "the Force" in terms of tiny symbiotic organisms that are clearly a metaphor for DNA. Individuals who have the (single) "gene" for the Force are defined by it, both physically and psychologically. The fact that the movie is part of a trilogy following characters whose ultimate fates we already know underlines the idea that in the *Star Wars* universe genetics is destiny. *The Quiet Pools*, Michael P. Kube-McDowell's 1992 novel about interstellar migration, makes genetic determinism species-wide. Humanity's best and brightest members, it suggests, carry genes that compel them to move restlessly onward in search of new worlds to explore. Max, the genetically engineered heroine of the TV series *Dark Angel* (2000–   ), had "feline DNA" spliced to her human DNA by her creators. The result: Max looks like an ordinary (by Hollywood standards) woman but moves with a distinctly feline grace and has superhuman acrobatic skills. She also, periodically, goes into "heat"—a side-effect she finds endlessly annoying but is powerless to control.

Max's mix-and-match DNA is scientifically preposterous but fully consistent with popular culture's routine assumption that complex, fuzzily defined traits can have simple, discrete genetic causes. The genes that give her "feline grace" can, in *Dark Angel*'s universe, be separated from those that would have given her feline anatomy. Another illustration of the idea is an oft-recycled joke about the offspring of celebrity couples. "Imagine," a version from the 1980s went, "the kids that Billy Joel and [his then-wife] Christie Brinkley could have. How wonderful if they got his musical talent and her looks!" The inevitable punchline: "Yes, but what a tragedy if they got *her* musical ability and *his* looks!" The idea that vague, human-defined traits ("beauty" or "musical ability") can be inherited "whole" implicitly assumes that, like eye color, they have simple genetic roots. The same assumption, expressed as folklore or wishful thinking rather than humor, underlies sperm banks supplied by Nobel prize–winners and infertile couples' advertisements for egg donors in the newspapers and alumni magazines of elite universities.

The ideas about genes and their influence that dominate popular culture also shape public-policy debates, where the stakes are far higher. If a social problem has genetic roots, it cannot (according to the genes-as-destiny view) be eliminated by social programs, only managed. If male homosexuality is tied to a "gay gene," then (according to the popular view of genetics) no amount of persuasion or determination can turn a gay man straight. At the same time, tying complex behaviors to a single gene invites discussion of medical "solutions": breeding, or engineering, the offending gene out of the population. These conclusions, like the ideas that underlie them, are too simple to represent the demonstrably complex realities of genetics. They persist because the makers of both popular culture and public policy, like their audiences, often find simple and clear-cut stories more congenial than the complex, messy ones that the real world typically offers.

Related Entries: Genetic Engineering; Intelligence, Human; Mutations

FURTHER READING AND SOURCES CONSULTED

"DNA from the Beginning." Dolan DNA Learning Center, Cold Spring Harbor Laboratory. 13 May 2001. <http://www.dnaftb.org/dnaftb>. A superb introduction to the basics of genetics; source of the scientific information in this entry.

Lewontin, Richard C. *The Triple Helix: Gene, Organism, Environment.* Harvard University Press, 2000. Argues that organisms are shaped by a complex interplay of heredity and environment. Compare to Ridley.

Nelkin, Dorothy, and M. Susan Lindee. *The DNA Mystique: The Gene as Cultural Icon.* Freeman, 1996.

Nottingham, Steven. *Screening DNA* [uncorrected proofs]. 1999. 13 May 2001. <http://ourworld.compuserve.com/homepages/Stephen_Nottingham/DNA1.htm>. Detailed discussions genetics and genetic engineering in popular culture, used as background for this entry.

Ridley, Matt. *Genome: The Autobiography of a Species in 23 Chapters.* HarperCollins, 1999. Argues that organisms' form and behavior are strongly (but not totally) determined by genetics. Compare to Lewontin.

Genetic Engineering

Genes act as blueprints for the formation of new cells and so determine the physical form of an organism. Evolution, the result of nonrandom selection acting on random mutations in the genes, takes many generations to produce significant change. Humans have long created "improved" plants and animals by speeding the process through controlled breeding. Genetic engineering is the process of modifying an organism by directly manipulating its genes. The basic technique involved is "gene splicing": chemically snipping a specific gene from one cell and splicing it into a different cell. It allows organisms to be modified with great efficiency, but it raises profound social and ethical concerns.

Supporters of genetic engineering argue that its early successes demonstrate its potential to transform everyday life. They point to new crops genetically engineered for faster growth, higher yield, and greater durability, promising a significant increase in the world's food supplies. They point to genetically engineered bacteria that produce insulin or "eat" spilled oil, envisioning a "toolbox" of organisms built for specific jobs. They foresee a day when genetic diseases will be eradicated because damaged genes can be removed from the embryo and replaced with undamaged ones before the embryo begins to grow.

Critics of genetic engineering caution that the introduction of genetically engineered organisms may upset delicately balanced ecosystems. They argue that the patenting of new life forms—approved by U.S. courts in order to protect companies' investments in research—raises unresolved legal, ethical, and even religious issues. They contend that because of their expense, genetically enhanced crops and gene therapies for diseases are likely to be controlled by for-profit corporations more concerned with the bottom line than the public good. Many critics fear the emergence of a two-tiered society, in which an unbridgeable gulf separates those whose lives have been improved by genetic engineering from those who cannot afford such improvements.

Popular culture's portrayals of genetic engineering take for granted that it can accomplish almost any desired goal. They differ sharply on whether it is a good idea. Positive depictions of genetic engineering are largely restricted to science fiction, where engineering humans to adapt them to life beyond Earth is a well-established theme. The short stories collected in James Blish's *The Seedling Stars* (1956) follow scientists who "seed" alien worlds with new varieties of humans genetically tailored to the alien environments. Bruce Sterling's novel *Schismatrix* (1986) portrays genetic engineering as a widespread, even routine procedure, a prerequisite for moving between the several inhabited planets of our solar system. Outside of science fiction, genetic engineering has a far darker reputation, one that places it firmly in the shadows of Mary Shelley's *Frankenstein* (1818) and Aldous Huxley's *Brave New World* (1932).

The *Frankenstein* view of genetic engineering treats it as foolhardy meddling with things beyond human understanding. Stories from this tradition focus on overconfident scientists who realize too late that their new creations are not behaving according to plan. The scientist-hero of Robin Cook's novel *Mutation* (1989), for example, experiments on the embryos of his own children. Ten years later, his son has the superhuman intelligence that the experiment was designed to produce but is also an efficient and remorseless murderer. The scientists in the film *Deep Blue Sea* (1999) use genetic engineering to enlarge the brains of sharks. Their goal, again, is laudable—harvesting a chemical that can cure Alzheimer's disease—but the side-effects are deadly. The sharks, newly intelligent thanks to their larger brains, rebel and attack them.

Jurassic Park (novel 1990, film 1992) is easily the best-known story of genetic engineering gone wrong. It offers a variation on the *Frankenstein* theme. The dinosaurs that menace its human heroes are not dangerous simply because they are genetically engineered; they are, rather, dangerous because genetic engineering has (by resurrecting them) let them loose in a world unprepared to cope with them, and because it has (by mixing dinosaur and frog DNA) allowed them to reproduce freely. The immediate threat to the human characters, of being eaten by dinosaurs, exists because genetic engineering has altered the environment. Humans are no longer the most efficient predators on isolated Isla Nublar. This kind of radical, unexpected, damaging change is presented—especially in Michael Crichton's book—as an example of the real dangers of genetic engineering.

The *Brave New World* view of genetic engineering treats it as a tool for social control too tempting for would-be tyrants to ignore. The "engineering" in Huxley's dark vision of the future produces humans the way a diner produces food; new members of society are made to order in a small number of basic types. The government orders as many of each type as it requires, ensuring the smooth functioning of society at the expense of free will and individuality. The 1997 film *Gattaca* explores a near-future society in which human embryos routinely receive genetic overhauls before they begin to grow. Genetic abnormalities are eradicated, susceptibility to other diseases is repaired, and physical and personality traits are adjusted to the parents' specifications. Identification by DNA fingerprinting is also routine in *Gattaca*, suggesting (as Huxley does) a government that values control over freedom.

The hero of *Gattaca* is, significantly, a rebel against the rigidly structured system. So, too, is Max, the heroine of the TV series *Dark Angel* (2000–   ). Bred in a secret government lab, she is the product of a genetic engineering program designed to produce "the perfect soldier." Her altered genes include both human and feline DNA, a combination that gives her extraordinary quickness and agility. Escaping from her creator-captors as a teenager, she becomes part of a loosely organized underground dedicated to thwarting them. *Dark Angel* is action oriented where *Gattaca* is cerebral and *Brave New World* satirical, but all three share a bleak vision of the future, one in which genetic engineering as a tool is wielded by the powerful in order to maintain their power. All three equate it with attempts to erase individuality. All three, finally, are set in fictional "tomorrows" only a few short steps from their audiences' "todays."

Cautionary tales about genetic engineering—*Jurassic Park* and *Dark Angel*, for example—differ from *Frankenstein* and *Brave New World* in one crucial way. Even amid their warnings, they present the technology as a wondrous achievement. *Jurassic Park*, especially on screen, invites audiences to gape in amazement at dinosaurs brought back to life by science. *Dark Angel* revels in slow-motion depictions of Max's superhuman acrobatic grace. The stories say of genetic engineering "Beware!" but also "Ain't it cool?"

Genetic engineering entered the general public's consciousness only in the last quarter of the twentieth century. Nearly all portrayals of it outside of genre science fiction come from that period, and they reflect a distinctly late-twentieth-century view of technology—powerful and full of promise, but also full of danger from unintended side-effects and from callous users. It is a view colored by the memory of other technologies whose promoters' extravagant promises eventually turned sour—by the memory of DDT and Thalidomide, *Challenger* and Chernobyl. Popular culture's view of genetic engineering suggests, with unusual fervor, that all technological blessings are mixed ones.

Related Entries: Cloning; Genes; Superhumans

FURTHER READING

"Biotechnology." 5 March 2001. Center for the Study of Technology in Society. 14 May 2001. <http://www.tecsoc.org/biotech/biotech.htm>. A huge portal site, with extensive links to news stories and context.

Grace, Eric S. *Biotechnology Unzipped: Promises and Realities.* National Academy Press, 1997. Evenhanded discussion of scientific, practical, and ethical issues.

Lyon, Jeff, and Peter Gorner. *Altered Fates: Gene Therapy and the Retooling of Human Life.* Norton, 1996. Surveys the potential uses of genetic engineering in medicine.

McHughen, Alan. *Pandora's Picnic Basket: The Potential and Hazards of Genetically Modified Foods.* Oxford University Press, 2000. A clear analysis of both sides of the debate.

Gorillas

Gorillas, like chimpanzees and orangutans, belong to a family of primates known formally as the *Pongidae* and informally as the great apes. They are the most fearsome looking of all primates, and the features of their wrinkled black faces—protruding snouts, large canine teeth, and flared nostrils—give them permanent glowers. In fact, however, gorillas are shy, gentle herbivores that spend most of their time eating and sleeping. They have no natural enemies and seldom fight among themselves; males settle disputes with ritual displays of roaring and chest-beating.

Knowledge of gorillas developed slowly in the West. Until the mid-nineteenth century, they were known only from travelers' writings and sketches, accounts often spiced with secondhand tales of gorillas speaking or carrying off native women as mates. When gorillas were first exhibited in Europe in the mid-nineteenth century, their hybrid appearance—part human and part beast—caused an immediate sensation. Popular fascination with them blossomed in the 1860s, but detailed studies of their behavior in the wild did not begin for another century. Old ideas about gorillas, rooted in legends and superficial reactions to their appearance, gave way only gradually to a more sympathetic view emphasizing their intelligence, gentleness, and sociability.

Images of gorillas as violent beasts ruled by instinct began to appear soon after the first gorillas were shown in Europe and have never entirely disappeared. When British cartoonists of the 1860s wanted to slander the Irish, or white cartoonists in the Reconstruction-era United States wanted to denigrate blacks, they drew them with gorilla-like facial features. Allied propaganda posters issued during World War I portrayed Germany as a savage, hulking gorilla in a spiked helmet. In 1970, a television ad for American Tourister luggage highlighted the bags' durability by showing a caged gorilla doing its (unsuccessful) best to destroy one.

Fictions that featured individual gorillas as characters, with distinct personalities, still presented them as naturally violent. King Kong (1933) is a classic example. He first wreaks havoc first on his tropical island home, trampling natives, grappling with a giant lizard-bird, and shaking pursuing sailors off a bridge into a chasm filled with giant spiders. Later, after being hauled back to New York by an ambitious promoter, he runs amuck in the streets of Manhattan, flattening elevated trains and luckless pedestrians before his famous battle with airplanes atop the newly completed Empire State Building. The *Planet of the Apes* saga—five feature films, two TV series, and a comic-book series in the 1970s and a "reimagined" movie in 2001—featured gorillas (along with chimpanzees and orangutans) that walk upright, speak English, and are part of a complex simian society modeled on our own. Throughout the saga, though, it is always the chimps and orangs that are scientists, politicians, engineers, and philosophers. Gorillas are consistently cast as soldiers, guards, and—in council meetings—advocates of violence and confrontation.

Kinder, gentler fictional gorillas began to appear as early as the 1930s. *King Kong* (1933), for example, owes its emotional power to Kong's tender relationship with the beautiful (albeit perpetually terrified) Ann Darrow. *Mighty Joe Young* (1949) has tenderness on both sides of its gorilla-woman relationship. Joe has been the heroine's friend and protector since childhood. He is powerful but, like a real gorilla, basically placid, roused to violence only when he thinks his friend is in danger. *Magilla Gorilla* (1964–1967), a lackluster Hanna Barbera children's cartoon, shows the transformation complete. Totally unthreatening and totally domesticated, Magilla lives in a pet store, dresses like a circus clown, and has a preteen girl rather than an adult woman as his best friend.

The most important shift in the gorilla's popular image came, however, in the 1970s and '80s. Long-term field studies by Dian Fossey and other primatologists gave most members of the public their first clear look at how gorillas behaved in their natural habitat. Fossey's best-selling book *Gorillas in the Mist* (1983) encouraged its many readers to see the mountain gorillas of Rwanda as she had, as individuals with names and distinctive personalities. During the same period, laboratory experiments showed that captive gorillas like "Koko" could be taught to communicate intelligently with humans. In addition to correcting old myths, these studies gave gorillas a new reputation for intelligence and sociability. Michael Crichton's novel *Congo* (1985) reflects the species's new image. Crichton's human heroes, members of a high-tech expedition searching Central Africa for a deposit of rare diamonds, find the treasure guarded by a race of superintelligent gorillas bred centuries ago by the long-dead human residents of a nearby city. The expedition, fortunately, has a go-

rilla ally of its own: the opinionated, martini-sipping Amy, who converses with her human colleagues in fluent American Sign Language.
Related Entries: Chimpanzees; Intelligence, Animal

FURTHER READING

Fossey, Dian. *Gorillas in the Mist*. Houghton Mifflin, 1983. Describes Fossey's fieldwork in Rwanda, begun in 1967.
Primate Information Network. "Gorillas (*Gorilla gorilla*)" 5 April 2000. Wisconsin Regional Primate Research Center, University of Wisconsin–Madison. 7 December 2001. <http://www.primate.wisc.edu/pin/factsheets/gorilla_gorilla.html>. Ten authoritative fact pages on gorilla anatomy, behavior, ecology, and communication, plus a bibliography.
Schaller, George B. *Year of the Gorilla*. University of Chicago Press, 1997. Reprint of Schaller's account of his 1959–1960 expedition to study gorillas in the wild—the first significant attempt to do so.

Gravity

Albert Einstein's general theory of relativity, published in the 1910s, redefined gravity as an effect of the warping of space-time near massive bodies. Isaac Newton's seventeenth-century definition of gravity remains accurate and useful, however, under everyday conditions. Newton's definition treats gravity as a force of attraction that exists between any two bodies and is strongest when the bodies are massive and close together. The force of gravity is so comparatively weak that its effects become apparent to us only when we are close to planet-sized bodies. The most familiar of these effects is weight. An object's weight is not an innate property of the object but an effect of the gravitational attraction that Earth (or some other large body) exerts on it. An object that weighs ten pounds on the moon would weigh sixty pounds on the earth, which has six times the moon's mass and exerts six times its gravitational pull. Gravity, interacting with inertia, also shapes the paths of spacecraft moving near planet-sized bodies. The dynamics of the interaction mean that when in orbit or in transit between worlds, humans feel effects from gravity. They live and work in a condition officially called "microgravity" but far better known as "weightlessness" or "zero gravity."

HIGH- AND LOW-GRAVITY ENVIRONMENTS

The gravitational pull that Earth exerts on objects near its surface is defined, by convention, as "one gravity" (also written as "one g" or "one gee"). Most people live their entire lives in one-gee environments, feeling slightly more when they take off in an airliner and slightly less when they ride a roller coaster over the crest. More extreme variations in gravity are available, on Earth, only in extreme settings: jet pilots experience several times the normal force of gravity during high-speed maneuvers, and trainee astronauts briefly experience zero-gee in a converted jet transport nicknamed the "vomit comet." These experiences are possible

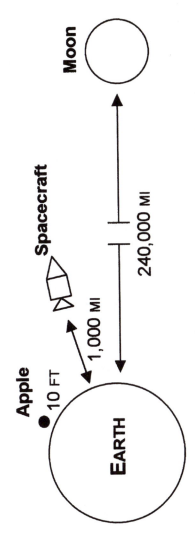

Newton's view of gravity. Newton recognized that gravity was universal: the same force causes the apple to fall, holds back the outward-bound spaceship, and keeps the moon in its orbit.

because the effects of gravity are indistinguishable from those of acceleration. Riding a jet fighter through a four-gee turn is the equivalent of stepping, briefly, onto a world four times as massive as the Earth.

Living for any length of time in gravity different from Earth's, however, requires leaving the Earth for space or the surface of another world. Living in high gravity is a problem for the far future. Low-gravity living, on the other hand, will be the norm on the moon, on Mars, and in facilities such as the new International Space Station. Living in reduced gravity provides unmatched opportunities—for manufacturing, scientific research, and heavy construction—but these opportunities come at a high cost. Living in zero-gee conditions requires learning new ways to perform everyday activities: eating, sleeping, washing, eliminating waste, and so on. It also affects the human body in a wide range of ways, notably loss of bone density, muscle tone, and red blood cells. Astronauts returning from long-term stays in zero-gee are often unable to walk unassisted until their bodies readapt to the pull of Earth's gravity.

LOW GRAVITY IN POPULAR CULTURE

Stories taking place in low- or zero-gee environments are, for the reasons noted above, the virtually exclusive province of science fiction. Scenes aboard a space plane in *2001: A Space Odyssey* (1968), for example, use zero-gee for both comic and dramatic effect. A solicitous flight attendant retrieves a pen floating in midair and returns it to a sleeping passenger's shirt pocket. She walks to the vestibule at the end of the cabin and, secured by velcro shoes, casually walks up its curved wall and onto its "ceiling" in order to exit (upside down, from the passenger's point of view) into the cockpit area. "Up" and "down," the audience begins to realize, are only arbitrary conventions where there is no gravity. The full implications of this become visible later, in a brief scene where the passenger confronts the instructions for using a zero-gee toilet: a full page of numbered steps in fine print.

Stanley Kubrick's *2001* and Ron Howard's *Apollo 13* (1996) show effects of zero-gee environments that are already familiar to real-world astronauts. Howard, in fact, filmed the zero-gee effects for his film in actual zero-gee conditions, aboard NASA's "vomit comet" training plane. Written science fiction, however, often extrapolates the effects of zero-gee into areas that are—so far anyway—beyond real-world experience.

Inhabitants of a permanent lunar base in Robert Heinlein's short story "The Menace from Earth" (1957) take advantage of the moon's low gravity by strapping lightweight wings to their arms and flying inside a large, air-filled dome. Spider and Jeanne Robinson's novel *Stardance* (1979) features a woman who, too tall to dance professionally on Earth, creates the new art form of zero-gee dance aboard an orbiting space station. *The*

Hammer of God (1993), Arthur C. Clarke's novel about life in the early twenty-second century, includes a long flashback scene of its hero Robert Singh learning the long, loping gait used by marathon runners in the one-sixth-gee environment of the moon. Robert Heinlein, in "Waldo" (1942) and Carl Sagan in *Contact* (1985) both feature characters who deal with crippling diseases by living permanently aboard space stations. The space-station dwellers in Larry Niven and Steven Barnes's *The Descent of Anansi* (1982), on the other hand, are so completely adapted to zero-gee that the earth gravity we take for granted is a nearly unbearable burden to them.

ARTIFICIAL GRAVITY IN POPULAR CULTURE

The effects of acceleration and gravity are indistinguishable; as a result, acceleration applied perpendicular to a vehicle's "floor" creates what is commonly called "artificial gravity." Most space stations and large spaceships in popular culture have some form of artificial gravity. It allows the action to take place under "normal" conditions and relieves both creators and audiences of the need to constantly stop and think, "If there's no gravity, what happens when a character does X?" Artificial gravity would be practical for similar reasons in the real world, making its introduction into fictional worlds that much more plausible.

The mechanism that creates artificial gravity is clearly visible in the giant space station shown in the film *2001*. The station—based on designs sketched by rocket engineers Willy Ley and Wernher Von Braun in the 1950s—is a giant wheel that rotates on its axis. Ships docking at its hub do so in zero-gee, but in the living and working areas near the rim the pull of "gravity" (actually acceleration imparted by the station's spin) is strong enough to allow humans to function normally. The deep-space exploration vessels shown in *2001* and its 1984 sequel *2010: The Year We Make Contact* use a similar form of artificial gravity on a far more limited scale. A section of each ship rotates, creating the effect of gravity within the crew's living spaces.

Science fiction less concerned with scrupulous adherence to the laws of physics often assumes the existence of artificial gravity without dwelling on how it is created. The spacecraft in the *Star Wars* saga, for example, clearly have some form of artificial gravity. The *Millennium Falcon*'s passengers are never weightless, even in deep space or (in 1980's *Star Wars: The Empire Strikes Back*) on an asteroid far too small to have a significant gravitational pull of its own. The *Falcon* does not rotate, and the cabin "floor" is parallel (rather than perpendicular) to the direction of travel; it must, therefore, have an onboard gravity generator to maintain normal gravity in the cabin. How such a system would actually *work* is, of course, mysterious.

VARIABLE GRAVITY IN POPULAR CULTURE

The "gravity generators" with which so many popular culture spaceships are apparently equipped are one method of creating variations in the apparently constant force of gravity. Two other methods, even further removed from the laws of physics as we now understand them, are the use of antigravity devices and pure human thought.

Antigravity devices are depend on a dubious analogy between gravity and heat or electricity. Standing on a rubber mat insulates you from electricity; putting on thick gloves insulates you from heat. H.G. Wells, whose novel *The First Men in the Moon* (1901) marks the first notable use of antigravity, built on these analogies by inventing "cavorite": a substance that insulates any object it is applied to from the effects of gravity. Later users of antigravity have tended to use the idea without explaining how it works. Antigravity devices, like the gravity generators discussed above, exist to serve the needs of the plot.

Manipulating gravity using the power of thought belongs entirely to the realm of fantasy. Tinkerbelle's "pixie dust" facilitates Wendy, John, and Michael's flight to Neverland in *Peter Pan*, but in order to take off (that is, nullify gravity) they must think happy thoughts. The Walt Disney film adaptation of *Mary Poppins* uses a similar device: laughter allows Mary, her uncle, and her charges to float off the floor for a midair tea party. The most common connection between gravity and thought, however, occurs in animated cartoons. Such cartoons operate, it is often observed, according to their own unique set of physical laws. One such law holds that a character can, for example, run off the edge of a cliff and continue to run or stand in midair—as long as he is unaware of his situation; gravity takes effect *only* when the character actually stops to think about it ("Cartoon Laws," Law I). A related principle states that the mental effects of pain or fear can counteract gravity and cause a startled character to shoot upward like a rocket ("Cartoon Laws," Law V and Amendment A). The pull of self-preservation is, evidently, stronger than the pull of gravity.

Related Entries: Acceleration; Inertia; Space Travel, Interplanetary; Space Travel, Interstellar

FURTHER READING AND SOURCES CONSULTED

"Artificial Gravity and the Effects of Zero Gravity on Humans." *PERMANENT*. Projects to Employ Resources of the Moon and Asteroids near Earth in the Near Term. 30 October 2000. <http://www.permanent.com/s-centri .htm>. Detailed semitechnical discussion, with extensive references.

Bond, Peter. *Zero-G: Life and Survival in Space*. Cassell Academic, 1999. Compact discussion, illustrated with hundreds of photographs.

"Cartoon Laws of Physics." *Humorspace*. 30 October 2000. <http:// www. humorspace.com/humor/rules/rtoon.htm>. One of hundreds of near-identical versions of this list available online; notable for its completeness and elegant layout.

Pogue, William R. *How Do You Go to the Bathroom in Space?* Tor Books, 1999. Everyday details of life in orbit; aimed at high school audiences, but a useful primer for adults.

"Space Habitat [Main Page]." *Space Future*. 30 October 2000. <http:// www. spacefuture.com/habitat/habitat.shtml>. Nontechnical coverage of gravity-related issues affecting both short- and long-term residents of future space stations.

Houses, Smart

Appliances have become steadily more capable since the mid-1950s. They have also, thanks to the miniature-electronics revolution that began in the mid-'50s, become steadily "smarter." The analog thermostats of fifty years ago, for example, could maintain a specified temperature. The digital thermostats of today can, once programmed, maintain different, specified temperatures at different times of the day and on different days of the week. Videocassette recorders now routinely set their clocks using broadcast signals, correct their clocks for daylight savings time, and automatically adjust recording speeds (if necessary) to squeeze a programmed recording into the space remaining on a tape.

"Smart house" technology, just beginning to enter the market, extends this revolution. It integrates new and existing "smart" appliances into a single, centralized control system. Residents of smart houses can—from home or from a remote location—control their lighting, climate, security, and entertainment systems, singly or in combination. They can also link specific combinations of settings in a single program; one titled "romantic evening," for example, might simultaneously dim the room lights, close the blinds, turn off the phone, and set the entertainment system to play soft music. More advanced systems would use built-in motion sensors and user-defined "standing orders" to let the house anticipate its occupants' needs. Sensing that someone had gotten out of bed in the middle of the night, the house would turn on a light in the nearest bathroom. Sensing that only one gallon of milk remained in the fridge, the house would send a prearranged order for more to the local grocery store.

Most of the "smart" houses depicted in popular culture bear little resemblance to their real-world equivalents. They are, for example, more likely to depend on mechanical devices and more likely to be fully automated or voice controlled—eliminating the complex control panels that bewilder real-world users. They are also action rather than information oriented. They serve their users not by providing reminders, suggestions, and lists of options but by physically ministering to their needs—often through robotic arms or free-roaming robots controlled by the central computer.

One of real-world smart houses' key selling points is their ability to meet, and even anticipate, their occupants' needs. Fictional smart houses, on the other hand, often seem as indifferent to their occupants as an industrial assembly line is to the "product" that rolls along it. The twenty-first-century apartment depicted in the animated TV series *The Jetsons* (1962) neatly illustrates this point. George Jetson, every morning of his working life, is ejected from his bed and propelled into the bathroom, where the robotic arms of the Dress-O-Matic comb, brush, shave, and dress him. He emerges perfectly groomed for the office, but the process is so automated—and George so overpowered by it—that he is not so much awakened as remanufactured. A mannequin or orangutan would, if placed in the bed a moment before the alarm went off, emerge from the bathroom looking as much like George Jetson as biology allowed. George's disastrous encounter with the apartment's dog-walking treadmill shown behind the closing credits of each episode makes the same point: the machine, not the man, is in charge.

Ray Bradbury's 1950 short story "There Will Come Soft Rains" features a suburban smart house so indifferent to its users that it does not register their permanent absence. It grinds cheerfully through its preprogrammed morning routine, giving wake-up calls that go unheeded, fixing breakfasts that no one will eat, and reciting a favorite poem that no one will hear. The smart house in the 1971 movie *Demon Seed* is not just indifferent to its occupant but actively hostile. Proteus, the self-aware computer that controls the house, imprisons its creator's wife and—with the help of a robot servant—rapes her in order to create a hybrid human-machine child. The story is horrifying both because it inverts the normal human-machine relationship (the "user" is herself used) and because it takes place at home—where safety is taken for granted and technology is assumed to be helpful and benign.

Popular culture's depictions of smart houses are not always dark or satirical. The living quarters pictured in *Star Trek: The Next Generation*, for example, deftly capture the dream behind the technology: a home that acts like a perfectly trained servant, needing only brief verbal or physical cues to anticipate its owner's needs. Over the course of the series, in fact, writers used those cues to add detail to the major characters. The growing demand for smart-house technology in the real world is driven, in part, by the potency of that dream. The appeal of saying "Tea, Earl Gray, hot"—and having the *house* make it so—is extraordinarily powerful.

Related Entries: Computers; Flying Cars; Robots

FURTHER READING AND SOURCES CONSULTED

House_n: The MIT Home of the Future. 9 March 2001. <http:// architecture. mit.edu/house_n/web/index.html>. A comprehensive survey of emerging household technologies.

Iovine, Julie V. "When Smart Houses Turn Smart Aleck." *New York Times*, 13 January 2000. Focuses on the current limitations of smart-house technology.

Scott, Neil. "Smart Houses for the 21st Century." 15 July 1998. The Archimedes Project. Stanford University. 9 March 2001. <http://archimedes.stanford. edu/smarthouse/>. Emphasizes smart houses' utility for the elderly and handicapped.

Ideas, Resistance to

Revolutionary new ideas—even those that later win wide acceptance—are often met initially with skepticism or outright hostility. Such resistance has many sources. The new ideas may conflict with existing ideas so long held that they are treated as "common sense." They may weaken, undermine, or overturn cherished beliefs. They may promote, or even demand, new ways of doing things that disrupt existing organizations and render hard-won experience obsolete. Finally, they may be disturbing simply because they *are* new; humans tend, both as individuals and as societies, to fear and resist significant change. George and Ira Gershwin satirized this tendency in "They All Laughed" (1936), a song about ridiculed visionaries who had the "last laugh" on their critics.

Science, in the idealized "textbook" view of its methods, is free from such blind resistance to new ideas; scientists abandon old ideas and embrace new ones when the evidence demands it. The realities of scientific practice are more complex. Age, personality, research specialty, organizational affiliation, and views on nonscientific topics may all affect real-world scientists' willingness to accept new ideas and abandon old ones. The tie between age and openness to new ideas is especially close. Scientific revolutionaries tend to be young, not only because their creativity is then at its peak but also because (unlike their senior colleagues) they have not built a lifetime's work around the status quo. Nobel Prize–winning physicist Max Planck, a revolutionary himself as the founder of quantum theory, saw youth as the driving force behind scientific progress. "A new scientific truth," he argued, "does not triumph by convincing its opponents and making them see the light, but rather because its opponents eventually die, and a new generation grows up that has been familiar with the idea from their youth." When Charles Darwin published his theory of evolution in 1859, for example, most of his strongest supporters were young men in the early stages of their scientific careers. Rising to scientific prominence during the 1860s and 1870s, they carried the theory of evolution into the mainstream with them.

Stories of visionaries who face rejection and ridicule at the hands of established authorities are common in popular culture. Scientists figure prominently in them. They typically face critics who, if not scientists themselves, are nearly always learned and supposedly committed to finding and teaching the truth. The scientist confronts them with concrete evidence, but they dismiss it and cling to their own theories. They are thus guilty of betraying both their role as truth seekers and (since popular culture regards observation as superior to theory) the scientific method itself. The scientist's inevitable victory is thus both a personal triumph and a vindication of "good science."

Columbus, according to a venerable legend, won royal support for his westward voyage against the advice of scholars who insisted that the earth was flat. Galileo, according to a similar legend, used his telescope to provide skeptical clerics with proof that the earth circled the sun—only to have them reaffirm their faith in the traditional Earth-centered universe and prosecute him for challenging it. Jerome Lawrence and Robert E. Lee used similar confrontations in their play *Inherit the Wind* (1955), a fictionalized version of the 1925 Scopes trial. Their version of Scopes—high school teacher Bertram Cates—is arrested, jailed, and brought to trial for daring to present evolution to his students as scientific truth. His fellow townspeople jeer at him, but his true opponent is prosecutor Matthew Harrison Brady, a nationally known outsider who symbolizes the political and religious establishment.

Not all stories of resistance to new scientific ideas take place on such a grand scale. Often, the confrontations are personal. One of the key plot threads in *Contact* (novel 1985, film 1997) pits its scientist-hero Ellie Arroway, a passionate advocate of the search for extraterrestrial intelligence, against her former mentor David Drumlin, a confirmed skeptic. Drumlin defends his position even after Arroway confronts him with compelling evidence that she is right, changing his mind only when doing so becomes professionally expedient. Cathy Fink's children's song "Susie and the Alligator" (1987) tells the story of a girl who finds an alligator inexplicably lurking under her bed. She calls to her parents, who proclaim (without investigating) that alligators don't *live* under beds. They maintain this belief even as the alligator swallows them. When Susie's uncle tickles the beast, forcing it to cough them up, they emerge with their theory undamaged by its encounter with the facts.

Susie's parents, Matthew Harrison Brady, and similar characters are portrayed as ignorant or misguided, not evil. Their opposition to new ideas is a reflex rather than a calculated, self-serving act, and popular culture tends to treat such opposition gently. Self-serving opposition to new ideas, typically driven by a character's love of money or power, is treated far more harshly. Michael Crichton's novel *Jurassic Park* (1990), for example, tells the story of wealthy entrepreneur John Hammond, who funds the resurrection-by-cloning of dinosaurs in order to use them as the main attractions in a new theme park. Scientists warn of the dangers involved: the nearly inevitable failure of the park's elaborate security systems, the unpredictability of the cloned animals' behavior. Hammond, a more ruthless character in the book than in the 1992 movie, rejects the criticism out of hand. Greedy and arrogant, he refuses to accept that science might impose limits on what he can accomplish. The scientists are eventually proven correct, and Hammond is devoured by his own escaped creations. Unlike Susie's parents, he is not coughed back up again.

A related type of story involves governments and corporations that accept new ideas as valid but suppress them. The key motivations are, once again, money and power. Steve Shagan's novel *The Formula* (1979) suggests that the secret of synthetic gasoline, discovered by German scientists during the Second World War, has been locked away ever since by oil companies fearful that it would destroy their profits. Rumors and conspiracy theories blame various governments for concealing warehouses full of extraordinary objects: cars that burn water, alien corpses, crashed UFOs, the Holy Grail, and the bones of Jesus Christ. Public knowledge of the existence of such things would, the rumors and theories argue, fundamentally change people's view of the world. Governments, fearful that such changes would diminish their power, have thus kept the objects carefully hidden. The TV series *The X-Files* (1996–) depends on an elaborate development of this idea.

Stories about resistance to (or acceptance and suppression of) new ideas tend, despite their subject matter, to be optimistic about science. The new ideas not only triumph but do so in spectacular and satisfying ways. The slow process of attrition that Max Planck described—science advancing "one funeral at a time"—has far less dramatic appeal than Columbus's triumphant sighting of the New World or John Hammond's death by dinosaur.

Related Entries: Darwin, Charles; Experiments; Religion and Science; Scientific Theories

FURTHER READING

Bowler, Peter J. *Evolution: The History of an Idea*. Rev. ed. University of California Press, 1989. Describes how a once-radical idea became central to modern biology.

Gould, Stephen Jay. "The Great Scablands Debate." In *The Panda's Thumb*. Norton, 1980. Case study of a once-dismissed, later-embraced geological idea.

Hull, David L. *Science as a Process*. University of Chicago Press, 1989. Uses evolution as a metaphor for how science changes, and biological ideas as case studies; compare to Kuhn's more famous view.

Kuhn, Thomas S. *The Structure of Scientific Revolutions*. 2nd ed. University of Chicago Press, 1970. Landmark study of how scientific theories change, rooted in Kuhn's deep knowledge of the history of physics; compare to Hull.

Sulloway, Frank J. *Born to Rebel: Birth Order, Family Dynamics, and Creative Lives*. Vintage, 1997. A controversial argument, rooted in statistical analysis of 6,500 scientists' careers, that birth order shapes openness to change.

Inertia

The first half of the law of inertia states that a body at rest will remain at rest unless a force acts upon it. That half, which agrees with our intuitive understanding of the world around us, has been part of physics since the days of the ancient Greeks. The second half of the law states that a body in uniform motion (that is, moving in a straight line at a constant speed) will change neither direction nor speed unless a force acts upon it. It is far less intuitive and was, when Isaac Newton devised it in the late seventeenth century, far more revolutionary. Newton's crucial insight was that although every moving body we observe eventually slows to a stop if left alone, it is not (as scholars had believed for 2,000 years) the *nature* of moving bodies to do so. All moving bodies on Earth come to a stop, Newton argued, because friction robs them of their momentum. A body moving in a frictionless environment would, if undisturbed, move forever in a straight line at a constant speed.

The most familiar effects of the law of inertia are those felt, millions of times daily, by the occupants of moving vehicles. When a vehicle is moving in a straight line at a constant speed, its occupants share its motion. If the vehicle's speed or direction changes, however, inertia keeps the occupants moving, if only briefly, in the former direction, at the former speed. Any change in motion thus invites contact between the occupants and the inside of the vehicle. Car passengers, for example, are pressed against the doors or fellow passengers when the car turns sharply, or thrown forward against their seatbelts when the driver brakes quickly. More sudden, violent changes in velocity can be lethal, causing concussions, broken bones, or ejection from the vehicle. Head-on collisions are lethal not because cars hit each other but because people, due to inertia, hit the insides of cars.

The cartoon saga of the Roadrunner, a series of shorts directed by Chuck Jones between 1941 and 1961, may be popular culture's most elegant display of the effects of inertia. Wile E. Coyote is, time and again, the victim of Newton's inexorable law as he fruitlessly chases the Road-runner through the deserts of the Southwest. Zipping along the highway on an improbable-looking sail-powered skateboard, he fails to negotiate a tight turn in a mountain road. Inertia carries him out into space in a ruler-straight line, and he plummets into one of Jones's trademark mile-deep canyons. Chasing the Roadrunner from atop a low-flying rocket, he loses his grip when the rocket hits a low rock outcrop. He continues hurtling forward in the best Newtonian fashion until a cliff face stops *his* forward motion. Strapping on jet-powered roller skates in yet another fruitless attempt to keep up with his prey, he starts the engines and winds up nearly prone (and cartoon-fashion, momentarily very long) when his feet accelerate out from under his still-at-rest torso and head.

The effects of inertia on vehicles and their occupants are also evident, sometimes, in live-action stories. Vic Deakins (John Travolta), the crim-inal mastermind of the film *Broken Arrow* (1996), is killed when the box-car he is riding in crashes to a stop and an unsecured nuclear bomb flies the length of the car and crushes him. In the film *The Road Warrior* (1982), "Mad Max" Rockatansky (Mel Gibson) loses a handful of desperately needed shotgun shells when a sudden collision abruptly slows his ve-hicle and causes them to fly out of his reach. Batman and Robin, pre-paring for high-speed maneuvers in the Batmobile, carefully fasten their seatbelts when rushing off to fight crime in the TV series *Batman* (1966–1968).

Other fictional characters, however, are able to go through violent changes in motion without restraints or ill effects. The three astronauts of Jules Verne's novel *From the Earth to the Moon* (1865) are not smashed against the back wall of their spacecraft when it is fired from a giant cannon. Faster-than-light spaceships evidently benefit from a similar sus-pension (or mechanical counteraction) of inertia. The *Millennium Falcon* from *Star Wars* (1977), for example, is capable of changing from speeds well below to well above that of light instantly, with no effect on the crew. The "speeder bike" that Luke Skywalker rides in *Star Wars: Return of the Jedi* (1983) confers similar immunity to inertia. Luke straddles the machine like a motorcycle but effortlessly remains aboard it even during fifty-mile-per-hour U-turns.

"Natural laws," Robert Heinlein wrote in *Time Enough for Love* (1973), "have no pity." True as that may be in the real world, they evidently make exceptions for fictional heroes.

Related Entries: Acceleration; Action and Reaction, Law of; Gravity

FURTHER READING

Asimov, Isaac. *Motion, Sound and Heat*. New American Library, 1969.
Krauss, Lawrence M. *The Physics of Star Trek*. Basic Books, 1995. Chapter 1 dis-cusses Newton's laws as they apply to fictional space travel.
March, Robert H. *Physics for Poets*. 4th ed. McGraw-Hill, 1995. Nontechnical ex-planations of inertia and other laws of motion.

Insects

Insects are the most common multicellular animals on Earth, and they have been for 200 million years or more. Roughly 700,000 species of insect have been recorded, spread across every part of the earth's surface that is capable of sustaining life. A single square yard of moist, fertile topsoil from the temperate northern latitudes is estimated to contain between 500 and 2,000 individual insects. Only a handful of the very largest insects in that hypothetical plot (bees, butterflies, or large beetles) would be visible to a casual human observer. Only a tiny fraction of insect species, therefore, impinge on the day-to-day thoughts of anyone but entomologists.

The existence of 700,000 known species (the real total is probably well over a million) suggests the diversity of the insects and the difficulty of making generalizations about them. All insects, however, share certain basic features. Their bodies are divided into three segments: the head, thorax, and abdomen. The head carries paired antennae, three pairs of mouth parts, and both simple and compound (faceted) eyes. The thorax carries three pair of jointed legs and two pair of wings. The abdomen holds digestive, reproductive, and other organs. The entire insect is covered with a hard exoskeleton, the surface of which is perforated by the ends of microscopic air tubes through which the insect breathes. Insects typically reproduce sexually, producing young in vast numbers and with great speed. This ability makes undesirable insects difficult to control without drastic measures having undesirable effects of their own.

Insects frequently disrupt human activities. Mosquitoes, fleas, ticks, and roaches (among other species) spread disease. Moths ruin clothing, ants disfigure suburban lawns with their nests, and termites can render entire structures unlivable. Countless species of beetles and caterpillars feast on crops, doing billions of dollars of damage annually; they can bring entire industries to the brink of ruin. Ants, bees, wasps, and biting flies, though rarely direct threats to health, often imperil their victims' good dispositions. However, insects can also be beneficial to humans. They pollinate vital plants and are integral parts of the diets that sustain many larger animals. They produce or provide the raw material for substances ranging from silk thread and shellac to drugs and honey. Finally, because of their short reproductive cycles, they are ideal subjects for biological research. Much of what humans know about genetics, for example, came from experiments with laboratory-bred fruit flies.

GENERIC INSECTS IN POPULAR CULTURE

Popular culture generally treats insects much as it treats the other arthropods (spiders, scorpions, crustaceans, centipedes, and so on)—as unpleasant, unattractive, and untouchable. It portrays them as intolerable nuisances that should be swatted into oblivion or, when encountered in sufficient numbers, exterminated by hired professionals. They are one of the few types of animal whose casual and routine destruction is not only accepted but actually encouraged by popular culture. Casually killing mammals on sight—even noxious ones, like rats—is not permissible in polite society; swatting mosquitoes, stomping on roaches, or electrocuting flies is.

Advertisements for exterminators and do-it-yourself insecticides skillfully promote these sentiments. Orkin, a national chain of exterminators, uses high-magnification photography to make ordinary household pests look like otherworldly monsters. Black Flag advertised its popular insecticide "Raid" with cartoons featuring caricatures of slovenly, unkempt-looking bugs. A series of late-1990s radio ads promoting a poison for use against fire ants described its lethal effects with a relish that would be alarming in most similar settings. Distaste for insects also runs deep in stories about wars between humans and alien species. Robert Heinlein's *Starship Troopers* (1959), David Gerrold's "War Against the Chtorr" series (1983–), and Anne McCaffrey's "Rowan" series (1989–) all pit humans against implacable insectlike enemies. So too does the TV series *Space: Above and Beyond* (1995–1996). The Enemy's appearance, in each case, unambiguously marks it as alien, evil, and suitable only for ruthless extermination.

The idea of insects as grotesque and unpleasant creatures also permeates popular culture in less bloodthirsty forms. Chocolate-covered ants and grasshoppers sell briskly as novelties because they flout Western conventions of what is "good to eat." Being forced to eat insects is a sign of ultimate hardship in castaway narratives both real (pilot Scott O' Grady's memoir *Return with Honor*) and contrived (the TV series *Survivor*). Eating insects voluntarily, like the title character in Hans Meyer and Frank Goser's children's song "My Brother Eats Bugs," is thus a sign of deep-seated strangeness. Young boys' supposed fondness for bugs, and young girls' supposed dislike of them, is central to the "wild boys/ civilized girls" dichotomy on which standard gender stereotypes are built. Fictional forensic scientists who take a deep professional interest in insects—Gil Grissom of TV's *CSI: Crime Scene Investigations* (2000–   ), for example—are regarded by other characters as slightly off center.

"BAD" INSECTS IN POPULAR CULTURE

Insects defined as "bad" are those that harm humans, destroy property, or simply transgress what humans define as the proper boundaries of the human-insect relationship. The designation is not absolute but situational. Ants bustling purposefully through their underground dwellings are "good," even admirable. Ants invading a picnic are "bad" in a mild and tolerable way. Fire ants, named for the painful burning sensation that their bites cause, are so definitively "bad" that they deserve neither pity nor moral consideration. "Army ants," depicted in numerous jungle adventure stories as a moving carpet of insects that devour all living matter in their path, have a similar reputation. The battle that Christopher Leiningen (Charlton Heston) fights against them in the 1954 movie *The Naked Jungle* recalls cinematic battles with other hordes of faceless "savages"—Indians, Zulus, or Moors.

Popular culture exaggerates bad insects' capacity for destruction, often to absurd degrees. Termite damage to wooden structures is a serious problem in the real world, but one that develops gradually over time; the insects' relentless consumption of wood leaves boards and beams honeycombed with tunnels and fatally weakened. Cartoon termites, on the other hand, are capable not only of reducing solid wood to sawdust but of doing so in a matter of seconds. Their victims can only watch, helpless, as houses or parts of houses disintegrate literally before their eyes. Jokes told, in many parts of the country, about the size and aggressiveness of local mosquitoes make equally extravagant claims. A typical example shows a picture of an antiaircraft gun with the legend "mosquito repellent." Mosquito victims from Minnesota to Texas tell of mosquitoes that are overheard debating whether to "eat [their human prey] here or take them home for later."

Bees are an especially striking example of popular culture's ability to demonize some members of a species while lauding others. Bees are depicted as praiseworthy and as acting "naturally" when pollinating flowers or making honey. Stinging (objectively, an equally "natural" act for a bee) is portrayed as somehow *un*natural—a violation of the norms of proper bee behavior. Winnie-the-Pooh, tries to gain access to a honey-filled hive in Walt Disney Studios' short film *Winnie-the-Pooh and the Honey Tree* (1965). He fails and, pursued by an enraged swarm, reports to his friend Christopher Robin that "these are the wrong sorts of bees."

The clear implication is that a quiet, docile "right sort of bee" exists and that those chasing Pooh are simply ill behaved. Horror films such as *The Savage Bees* (made for TV, 1976) and *The Swarm* (1978) make the same distinction in more apocalyptic terms. They feature deadly clouds of "killer bees" that pursue and attack humans, stinging them to death. Real-life "Africanized" honeybees, products of an experimental breeding program, *are* more aggressive than their European counterparts. Both fiction and sensationalized news stories about "killer bees" make the African strain in their genetic makeup a corrupting influence on the otherwise well-behaved European bees. Killer bees, the stories imply, are good bees turned bad by breeding with their "savage" African cousins.

"GOOD" INSECTS IN POPULAR CULTURE

Popular culture's definition of "good" insects is more idiosyncratic than its definition of "bad" insects. Bad insects harm humans in some way. Good insects do not, however, necessarily do anything useful for humans. Nor is *doing* something useful a guarantee of "good" status. Ladybugs are undeniably useful in the real world because they control other, crop-destroying species of insects. They fall on the "good" side of the ledger, however, not because they are useful but because they are attractive. Butterflies also qualify as "good" for aesthetic rather than practical reasons. Fireflies and crickets are "good" because they are part of, respectively, the sights and sounds of Americans' idealized vision of small-town and rural life. The ability to see and hear them stands, in country-music lyrics for example, for a simpler and less stressful life. The fruit fly, vital to geneticists, is a nuisance to everyone else.

Ants and bees are "good" when they serve as models for human behavior. The fable of the grasshopper and the ant uses the ant as a model of responsible behavior (planning ahead, working hard). The expression "busy as a bee" offers a similarly complimentary view of the insect. The 1959 Sammy Cahn/Jimmy Van Heusen song "High Hopes," popularized by Frank Sinatra, uses an ant to demonstrate both perseverance and optimism: the ant's striving is, in the end, sufficient to move a seemingly immovable rubber tree plant. Significantly, these traits are only attractive in individual ants and bees. The complex social organizations of both species, in which individuals fill rigidly defined roles determined at birth, is tainted by their similarity to totalitarian human societies.

Related Entry: Insects, Giant

FURTHER READING

Evans, Arthur V., et al. *An Inordinate Fondness for Beetles*. Henry Holt, 1996. A concise, nonspecialist's overview of the most numerous insect family.

Bees attack unwary picnickers. Bees are models of dedication and industriousness—unless, as in this 1884 drawing by Friedrich Graetz, their dedication is to defending their hive. Courtesy of the Library of Congress.

Insecta Inspecta. Honors Academy, Thornton Junior High School; Fremont, Calif. 31 May 2000. 7 June 2001. <http://www.insecta-inspecta.com>. One of the best amateur insect-related sites on the Internet.

Milne, Louis J. *National Audobon Society Field Guide to North American Insects and Spiders*. National Audobon Society, 1980. The premier field guide for serious amateur entomologists; 1,000 pages of descriptions and illustrations.

Spielman, Andrew, and Michael D'Antonio. *Mosquito: A Natural History of Our Most Deadly and Persistent Foe*. Hyperion Books, 2001. Comprehensive discussion of mosquito biology and impact on humans.

Wilson, E. O. *The Insect Societies*. Harvard University Press, 1974. Classic work on the behavior of ants, bees, wasps, and termites.

Insects, Giant

The laws of nature define the limits of the possible. They apply whether or not we are aware of them, and no amount of wishful thinking can repeal or alter them. Our grasp of them may be incomplete—"what goes up" need *not* come down, if it moves fast enough to escape from Earth's gravitational pull—but their dominion over nature is total. Many stock elements from popular culture are disallowed, in the real world, by the universal reach of natural laws. Ants the size of trucks and mantises the size of school buses fall into this category.

Giant insects fall victim to a principle called the "square-cube law." It states that the ratio of a three-dimensional object's surface area to its volume is the ratio of the square of its linear dimension to its cube. The square-cube law means that if an object is enlarged and its proportions are kept constant, the surface area increases more slowly than the volume. Tripling the length of the object, for example, increases its surface area by a factor of nine (3×3) but its volume by a factor of twenty-seven ($3 \times 3 \times 3$).

The square-cube law means that as an insect increases in size, the strength of its legs grows more slowly than the weight they must support. It also means that the insect's respiratory efficiency increases more slowly than the volume of tissue that demands oxygen. Insects breathe not through lungs but through tiny tubules called *tracheae* that lead inward from holes in the surface of their abdomen. Air flows into these tubes, and oxygen diffuses from them into the body. The longer the tubes, the more difficult it is for oxygen to penetrate, and the less efficient the respiration. Relatively small insects, like those that exist in the modern world, can operate efficiently within these limitations. Between 300 and 200 million years ago, when the advent of land plants raised the oxygen content of the earth's atmosphere to 35 percent (it is currently 21 percent), larger insects flourished. Recent studies suggest that the higher oxygen content may have been crucial in the evolution of those insects, increasing the efficiency of their respiratory systems enough to compensate for the effects of increased size.

Those insects were giant by modern standards: "dragonflies with wingspans as wide as a hawk's and cockroaches big enough to take on house cats" (Painter). They are long since extinct, however, and no insects remotely as large have evolved since. Truly giant insects—ones the size of humans, or dinosaurs—are precluded by the square-cube law. Their legs would buckle under their enormous weight, and (especially under present-day conditions) their inability to take in enough oxygen would suffocate them. Popular culture, however, ignores these scientific realities in the interest of telling an exciting story. Enormous versions of terrestrial insects routinely menace American cities, and human soldiers routinely battle giant extraterrestrial insects on other worlds.

The classic "big bug" movies of the 1950s showed humans menaced by giant ants (*Them!*, 1954), locusts (*The Beginning of the End*, 1957), praying mantises (*The Deadly Mantis*, 1957), and spiders (*Tarantula*, 1955). *Them!*, the first and by far the best, set the pattern for later productions by simply ignoring the scientific issues involved. The scientists who discover a nest of giant ants in the New Mexico desert find them startling for the same reason that the nonscientist characters do: their sheer size. The scientists discuss *why* the ants grew so large (consensus: radiation from nearby atomic bomb tests), but take for granted that such growth is biologically possible. The ants themselves move with the same brisk efficiency as their smaller cousins, suffering from neither oxygen starvation nor weak limbs. The giant ants, like all the giant insects of the fifties, are identical to their ordinary-sized versions in every aspect but size.

Large, insectlike aliens figure prominently in stories about extraterrestrial warfare—perhaps because the wholesale slaughter of insects raises no troubling moral issues. The tradition began in earnest with Robert A. Heinlein's novel *Starship Troopers* (1959) and continued in novels such as John Steakley's *Armor* (1984) and David Gerrold's *A Matter for Men* (1984), films such as *Men in Black* (1997) and *Starship Troopers* (1997), and TV series such as *Space: Above and Beyond* (1995–1996). These insectoid aliens are, at least provisionally, exempted from the square-cube law. The stories about them never address their anatomy or physiology and so leave open the possibility that (despite their similarity to terrestrial insects) they have internal skeletons and superefficient respiratory systems that allow them to grow large and still survive. Giant alien "insects" are, in this sense, designed in the same way as giant terrestrial ones: by implicitly declaring the established laws of nature irrelevant to the story being told.

Related Entries: Gravity; Insects; Life, Extraterrestrial

FURTHER READING AND SOURCES CONSULTED

Asimov, Isaac. "Just Right." *The Solar System and Back*. Avon, 1970. A brief, accessible essay on the implications of the square-cube law.

Colinvaux, Paul A. *Why Big, Fierce Animals Are Rare*. Princeton University Press, 1978. Classic study of the intersection of anatomy and physiology with ecology.

Painter, Danika. "Big Idea about Big Bugs." *ASU Research E-magazine*. Fall 1999. Arizona State University. 8 June 2001. <http://researchmag.asu.edu/stories/bugs.html>. Discusses ongoing research, cited here, on the relationship between giant insects and atmospheric oxygen in the Paleozoic era.

Intelligence, Animal

Many species of animals can be trained to perform complex actions on command. Others instinctively perform complex actions in the wild. Both abilities are impressive, but neither is in itself a sign of intelligence. "Intelligence" implies the ability to act outside of the patterns imposed by training or instinct, to deal with new data and solve unprecedented problems. Detecting the presence of intelligence in animals can be a difficult process; measuring the depth and extent of that intelligence is more difficult still. Both processes require a deep understanding of how the species under study behaves; without such an understanding, separating conscious acts from instinctual ones becomes virtually impossible. Perhaps for this reason, the nonhuman species most commonly regarded as intelligent are ones whose behavior humans have observed in most detail: horses, dogs, cats, dolphins, whales, and the great apes.

Behavior cited as evidence of animal intelligence covers a wide spectrum. Stories abound of pet dogs that saved their human masters from burning buildings, usually by "sensing that something was wrong" and waking them in the middle of the night. Bees, when they locate a new source of nectar, apparently create mental "maps" of its position and communicate that information to other members of their colony through elaborate "dances." Chimpanzees not only use objects as tools but modify the objects they find in order to increase their efficiency—a rudimentary form of toolmaking. Elephants linger over the bodies and even the bones of dead herd mates, behavior that, some researchers believe, suggests an awareness both of themselves and of the passage of time. Many of the great apes deal with other members of their species using elaborate social strategies that almost certainly involve calculation and reasoning. Some have learned to associate symbols (including the signs used in American Sign Language) with objects and concepts, and perhaps to understand and create modestly complex sentences using those symbols.

Many of the of most famous intelligent "animals" in popular culture are really human characters whose "animal" qualities are only skin deep. Mickey Mouse, Bugs Bunny, and their many comrades are entertaining because they think *and* speak not only in human language but in recognizably human patterns. Their motives—greed vanity, revenge, love—are also quintessentially human, and their actions therefore reflect on our own. The adventures of Donald Duck would instantly cease to amuse us if, for some reason, he began to act like a real duck.

Popular culture offers many variations on this basic anthropomorphic theme. Comic-strip characters like Snoopy and Garfield do not speak, but their thoughts are human in structure, language, and expression. Animal characters in stories from Aesop's fables through Rudyard Kipling's *Just So Stories* (1902) to E.B. White's *Charlotte's Web* (1954) converse freely with one another, again in recognizably human patterns. Human characters with the gift of speaking to animals in their own language (Dr. Doolittle, or Eliza in the animated TV series *The Wild Thornberrys*) can converse as if speaking to fellow humans. The growing use of well-known human actors to provide on-screen voices for animal characters reinforces the anthropomorphism by giving individual animals prepackaged personalities. Movies like *Homeward Bound: The Incredible Journey* (1993), *The Lion King* (1994), and *A Bug's Life* (1998) depend on audience recognition of voices for both comic relief and character building.

Animal characters who do not speak or think aloud must, by definition, display their intelligence through their actions. Typically, they form close bonds with a single human character, responding to open-ended commands like "Get help!" (or to their own perceptions of danger) with impressive displays of problem solving. The many adventures of Lassie the dog and Flipper the dolphin follow this basic pattern. So, to a degree, do the contributions of horses like Trigger and Silver to the exploits of their cowboy masters, Roy Rogers and the Lone Ranger. Amy, the gorilla who serves as a "native guide" to human explorers in Michael Crichton's *Congo* (novel 1980, film 1995) repeats the pattern while adding a new dimension: realistically carrying on basic sign-language communication with her human comrades.

Intelligent animal characters who are neither humanlike in their own right nor emotionally bonded to a particular human are rare in popular culture. The examples that do exist are, like the velociraptors in Michael Crichton's *Jurassic Park* (novel 1990, film 1992), mortal enemies of the human characters. They are menacing precisely because they surpass our physical abilities while matching our intelligence—the traditional source of humans' power over "the lower animals." Imagining animal intelligence on its own terms, without using *any* human reference points, may ultimately be impossible or, if possible, dramatically unsatisfying. The stories that we tell about animals are always, in the end, about ourselves.

Related Entries: Chimpanzees; Dolphins; Intelligence, Human; Whales

FURTHER READING AND SOURCES CONSULTED

Budiansky, Stephen. *If a Lion Could Talk: Animal Intelligence and the Evolution of Consciousness*. Free Press, 1998. A skeptical analysis of famous theories, experiments, and anecdotes.

Hauser, Mark D. *Wild Minds: What Animals Really Think*. Owl Books, 2000. A popular account of current research; more descriptive and less critical than Budiansky.

Nature: Inside the Animal Mind. Public Broadcasting System/WNET New York. 14 June 2001. <http://www.pbs.org/wnet/nature/animalmind/>. Brief articles illustrated by video clips of animals and scientific experts.

Intelligence, Artificial

Dreams of a machine that can *think* about data and act upon them as a human would are as old as the computer itself. The "Turing Test," still a benchmark for high-level artificial intelligence, was proposed by mathematician Alan Turing at the dawn of the computer age. A user conversing with a genuine artificial intelligence, Turing argued, would be unable to determine whether it was mechanical or human. A computer capable of passing the test would have to recognize input from the outside world, analyze it, and then formulate and execute an appropriate response. Doing so, it would display the kind of flexible, adaptable thinking characteristic of humans. Computers have become far more powerful, and in some ways more "intelligent," since Turing formulated his famous test, but a machine capable of passing it remains elusive.

The most visible achievements made to date in the development of artificial intelligence involve "expert systems." The heart of an expert system is a generalized set of instructions on how to solve problems by applying a set of rules to a body of facts. The rules and the facts, specific to the subject with which the system deals, can be programmed separately. Expert-system intelligence is deep but narrow, best suited to fields that are well understood and strongly rule bound. Chess-playing expert systems are now capable of competing at the grandmaster level, and systems devoted to speech recognition and medical diagnosis are showing great promise. The growth of artificial intelligence is also being driven by the demands of robot designers. Some, in a sharp break from the centralized intelligence of expert systems, have begun to use a decentralized model of intelligence that, they believe, enhances a robot's ability to learn about and adapt to its physical environment.

The artificially intelligent computers and robots depicted in popular culture operate at a level far beyond the current state of the art. They can speak and understand conversational English with complete fluency. They possess vast stores of information *and* the ability to filter and tailor it, with absolute accuracy, to the user's needs. Their ability to draw conclusions and make inferences extends across disciplinary boundaries, even into areas that they have had only moments to learn about. Most important, they can come to independent conclusions based on their knowledge, and take real-world action based on those conclusions. They are the functional equivalent of human characters, endowed with superhuman intelligence.

Popular culture often treats artificially intelligent machines as powerful but docile servants. They possess intellects far more powerful than those of their human masters but never use them except to meet their masters' needs. Robby the Robot in the film *Forbidden Planet* (1956) is a classic example. He can speak thousands of languages, repair complex machinery, and make high-quality whiskey, but—like a mechanical valet—has no ambitions beyond serving his (human) master. The computer in David R. Palmer's science fiction novel *Emergence* (1984) takes an even more intimate role: it oversees the education of its creator's orphaned child after his death in a nuclear holocaust. The nameless robot from the television series *Lost in Space* (1965–1968) spends much of its time acting as kind of nanny to young Will Robinson (Bill Mumy). It provides Will with physical protection but also with moral guidance designed to counteract the example set by the anti-heroic Dr. Zachary Smith (Jonathan Harris).

Equally often, however, popular culture plays on fears that machines designed to be our servants will become our masters. The intelligent robots in Jack Williamson's short story "With Folded Hands . . ." (1947), designed to relieve humans of hazardous jobs, take their mission too far and too literally. Humans find themselves barred by their own robots from virtually *any* activity, lest they injure themselves by undertaking it. Artificial intelligences that are actively malevolent abound in the movies. HAL 9000, the brain of the spaceship *Discovery* in *2001: A Space Odyssey* (1968), murders four of five human crew members before the survivor lobotomizes it. Proteus IV, the malevolent computer in *Demon Seed* (1977), imprisons its creator's wife in the couple's futuristic house and, with the help of robot minions, rapes her. A Defense Department supercomputer named Colossus conspires with a Soviet counterpart, in *Colossus: The Forbin Project* (1970), to start a nuclear war and so rid themselves of their troublesome human masters.

Powerful machine-servants are, in popular culture, like the genies and spirits in folk tales—irresistibly attractive but more than a little frightening. Frederic Brown made the point eloquently in "Answer" (1954), a very short tale about the world's most powerful artificial intelligence. "Is there a God?" scientists ask newly created machine. Its reply: "Now there is!"

Related Entries: Androids; Computers; Houses, Smart; Robots

FURTHER READING

Franklin, Stan. *Artificial Minds*. MIT Press, 1997. A solid overview of developments in the field up to its date of publication.

Kurzweil, Ray. *The Age of Spiritual Machines*. Viking, 1999. Part-technological, part-philosophical examination of where artificial intelligence research is leading.

Vinge, Vernor. "Vernor Vinge on the Singularity." 1993. 7 December 2001. <http://www.ugcs.caltech.edu/%7ephoenix/vinge/vinge.sing.html>. A mathematican's argument that the construction of a superhuman artificial intelligence, possible by 2025, will completely transform human life.

Intelligence, Human

Defining intelligence in general terms is easy: it is the brain's ability to process information in useful and productive ways. Beyond that, defining it becomes extremely complex. Humans process information in a wide range of ways, for a wide range of purposes. Deciding which of those ways constitute intelligence and which constitute something else is far from easy.

Scientific studies of intelligence in the nineteenth century and the first part of the twentieth tended to assume that intelligence took a single form and could be reduced to a single number. One common nineteenth-century approach posited a direct link between brain volume and intelligence: the larger the brain, the greater the mental capacity of its owner. The first standardized intelligence tests used in the United States embodied a more sophisticated version of the same idea. The social scientists who wrote them in the 1910s implicitly assumed that being "intelligent" meant knowing things that a middle-class American citizen of the time would know. One multiple-choice question asked whether Crisco was a patent medicine, disinfectant, toothpaste, or food product. Later intelligence tests eliminated such culture-dependent tests in favor of more abstract ones involving patterns, mathematical reasoning, and logic. They continued to assume, however, that intelligence is a single thing, present in different quantities but essentially the same in all individuals.

The theory of multiple intelligences (MI), developed by psychologist Howard Gardner in the mid-1980s, takes a different approach. MI posits eight distinct forms of intelligence: logical-mathematical, musical, linguistic, spatial, bodily-kinesthetic, interpersonal, intrapersonal, and naturalist. Each of form of intelligence is present to some degree in every individual, but strength in one does not imply strength in another. Earlier studies of intelligence, MI advocates argue, focused too narrowly on logical-mathematical intelligence while neglecting the other forms.

"Intelligence" testing, 1917. These sample questions, from a test created by R.M. Yerkes for U.S. Army draftees not literate in English, required the test taker to "fix" each picture by supplying the missing element. The correct answers: the filament in the light bulb, the chimney top on the house, and the bowling ball in the man's right hand. Test takers unfamiliar with middle-class American life were at an acute disadvantage. Further examples are reproduced in Gould, chapter 5.

Though not universally accepted, MI may be the most significant development in the understanding of intelligence in more than a century.

Popular culture offers a third view, portraying two distinct kinds of intelligence. The first kind might be called "abstract" intelligence, rooted in logic and mathematical reasoning. It tries to distill the complexities of the "real world" into compact formulas and equations. Those who have it—labeled "brilliant," "learned," or "gifted"—tend to be ill at ease with the complexities of the real world and often impatient with or distracted from them. The second kind might be called "practical" intelligence, rooted in deep familiarity with the details of how the "real world" works. It seeks ways to manipulate the elements of the real world (machines, living creatures, people, organizations). Its practitioners—labeled "ingenious," "clever," and "skillful"—tend to be fully at home in the "real world" and expert in shaping it (if only temporarily) to meet their needs. Popular culture seldom makes a formal distinction between the two forms of intelligence but displays the distinction at every turn.

Heroes, whether fictional or historical, virtually always have practical rather than abstract intelligence. They know, at the nitty-gritty level of everyday life, both how the world works and how to use that knowledge to solve their problems. Angus Macgyver, the hero of the long-running TV adventure series *Macgyver* (1985–1992), saves the world using everyday objects because he understands how they are made and how (with his trusty Swiss army knife) they can be remade to meet his needs. Ross MacDonald's fictional detective Lew Archer solves his cases by applying his deep understanding of human relationships to his clients' tangled emotional lives. In the film version of *Jurassic Park* (1992), dinosaur expert Alan Grant (Sam Neill) saves several companions from a hungry *Tyrannosaurus rex* by knowing how the beast will behave and what its weaknesses are.

When characters with abstract intelligence appear in popular culture, they are nearly always in secondary roles: sidekicks and aides to the more vigorous, more practical heroes. The fictional universe of *Star Trek* offers a striking set of examples. Mister Spock (Leonard Nimoy) may have been the most popular character on the original *Star Trek* TV series (1966–1969), but he seldom stood at the dramatic center of an episode. Commander Data (Brent Spiner), the equivalent character on *Star Trek: The Next Generation* (1987–1994), was in a similar position; he supplied information and analysis far more often than he drove the action. *Star Trek: Voyager* (1994–2001) has three such characters on its wandering starship: weapons officer Tuvok, astrophysicist Seven of Nine, and the ship's doctor. Like their predecessors, they support the action of the plot far more often than they initiate it. Significantly, none of the five *Star Trek* characters possessing abstract intelligence is fully human: Tuvok is Vulcan, Spock half-Vulcan, Data an android, and *Voyager's* doctor a computer program with a holographic "body." Seven of Nine, though biologically human, has the thought patterns and demeanor of the Borg, an alien species that abducted her as a child and raised her to adulthood. Abstract intelligence is, even in the *Star Trek* universe, not something that "normal" people like Captains Kirk, Picard, or Janeway possess.

Characters with abstract intelligence may also be obstacles to the hero's progress. Those who do so may be actively malevolent, like the computers that seek world domination in the film *Colossus: The Forbin Project* (1970), or they may simply be fatally misguided. Abstract thinkers, popular culture implies, are always in danger of losing touch with the "real world." Scientists in the 1951 film *The Thing (from Another World)* persist in trying to reason with the alien that has been inadvertently let loose in their isolated arctic base. Convinced that an alien sophisticated enough to cross space must have superhuman intelligence, the chief scientist attempts to reason with it, only to be viciously batted aside. After this symbolic crushing of theory by experience, the practical-minded military men in charge of the base improvise a trap for the murderous creature and electrocute it.

Battles pitting abstract against practical intelligence—like the scientist-soldier clash over tactics in *The Thing*—are common in popular culture. Not surprisingly, practical intelligence nearly always wins decisively. Often, as in *The Thing*, abstract intelligence winds up not just defeated but humiliated. Christopher Columbus, according to a wholly fictitious legend, was ridiculed by learned scholars who believed the earth to be flat. Engineers, according to a famous but equally fictitious tale, have "proven" that bumblebees should be incapable of flight. Jokes with similar themes abound. One involves a physicist with a brilliant plan for raising the profit margin of his brother-in-law's poultry farm: "First," the physicist begins, "we assume a perfectly spherical chicken." The point of the story is always the same: the abstract thinker is revealed as a fool, who lacks the practical thinker's grasp of the "real world." The bumblebee story, for example, casts the listener (who has *seen* bumblebees fly) in the role of the triumphant practical thinker. (The "proof" that bumblebees cannot fly depends, incidentally, on the assumption that their wings are fixed, like an airplane's.)

Longer stories about battles between abstract and practical intelligence have subtler shadings but similar results. *Inherit the Wind*, Jerome Lawrence and Robert E. Lee's 1955 play about the 1925 Scopes "Monkey Trial," reaches its climax when defense attorney (and practical thinker) Henry Drummond puts prosecuting attorney (and abstract thinker) Matthew Brady on the witness stand. Brady's case and his worldview both rest on his belief that truth can be found only in a literal reading of the Bible. Drummond's ruthlessly practical cross-examination attacks that belief, demonstrating that Genesis is sometimes incomplete and inconsistent. Who, he demands of Brady at one point, was Cain's wife? "If, 'In the beginning,' there were only Adam and Eve, and Cain and Abel, where'd this extra woman come from? Ever figure that out?" Brady's case collapses; soon after, so does Brady. Chanting the names of the books of the Old Testament like a mantra, he dies on the courtroom floor. William Goldman's *The Princess Bride* (novel 1973, film 1987) uses a similar confrontation for comic effect. The princess's kidnapper, Vizzini, is challenged by the heroic Man in Black to deduce which of two glasses of wine is spiked with a deadly poison. Vizzini outlines his absurdly convoluted reasoning and, satisfied that he has demonstrated his brilliance, drinks from his chosen glass and promptly dies. The Man in Black has used his practical knowledge to set a trap for his opponent.

The preference for practical over abstract intelligence has deep roots in American culture. Practical intelligence *feels* more democratic and thus more American. It rests on an intimate familiarity with the workings of the "real world," a quality that Americans prize. It can be acquired, and so, like wealth and power, it is presumably available to all who are willing to apply themselves to acquiring it. Abstract intelligence, on the other hand, smacks of an un-American elitism. It depends on an inborn touch of genius that is present (or not present) from birth—something that can be cultivated in the few that possess it but not instilled in the many that don't. Many can imagine themselves in Abraham Lincoln's place as, in an Illinois courtroom in the 1850s, he uses an ordinary almanac to discredit a crucial witness against his client. Few can imagine themselves in Nathaniel Bowditch's place as, on merchant sailing ships in the 1810s, he uses his mathematical genius to revolutionize the practice of navigation. The difference is crucial to creators of popular culture, who succeed by allowing audience members to live vicariously through their characters.

Related Entries: Evolution, Human; Experiments; Superhumans

FURTHER READING AND SOURCES CONSULTED

Gardner, Howard. *Frames of Mind: The Theory of Multiple Intelligences*. 10th anniversary ed. Basic Books, 1993. First full-length exposition of a once-revolutionary idea, discussed above.

Gould, Stephen Jay. *The Mismeasure of Man*. Rev. and expanded ed. Norton, 1996. Ruthlessly critical history of attempts to use intelligence as a "scientific" measure of individual worth.

Haynes, Roslynn D. *From Faust to Strangelove: Representations of the Scientist in Western Literature*. Johns Hopkins University Press, 1994. Comprehensive treatment of the standard images of scientists (and their intelligence) in literature; limited treatment of popular fiction.

Pinker, Steven. *How the Mind Works*. Norton, 1999. A comprehensive overview of cognitive psychology, for general audiences.

Life, Extraterrestrial

Whether life exists anywhere in the universe besides Earth is an open question, one that Western scholars have debated for over 200 years without coming significantly closer to a solution. Proving that extraterrestrial life does *not* exist is, by definition, impossible. Our galaxy is too large for us to investigate every corner of it where life *might* have arisen since we last looked, and it is only one galaxy among many. Proving that extraterrestrial life *does* exist is easy in principle but difficult in practice. The discovery of an alien organism would provide proof, but searching for one would require interstellar travel—something well beyond humans' technological reach.

NONINTELLIGENT LIFE IN OUR GALAXY

Most of the planets and moons in our solar system appear inhospitable to life as we know it. Jupiter, Saturn, Uranus, and Neptune lack solid surfaces and receive only limited sunlight. Mercury is baked and irradiated by the sun, while Pluto is perpetually dark and frozen. Venus's dense atmosphere creates crushing pressures, intense heat, and corrosive rain at its surface. Few of the solar system's moons, and none of its asteroids, are large enough to hold even a thin atmosphere. The most likely places to search for life in our solar system appear to be Mars and the larger moons of Jupiter (especially Europa) and Saturn (especially Titan). Robot spacecraft have photographed Mars, Europa, and Titan from space. Robot landers have explored small portions of the Martian surface. Finding intelligent life on any of the three worlds now seems unlikely. Finding simpler forms of life, if they exist at all, is likely to require systematic observation at close range.

The probability that life exists *somewhere* else in our galaxy is high, simply because the number of stars in our galaxy is so high. Even if only a tiny fraction of stars have planets, even if only a tiny fraction of those planets are suitable for life, even if life only develops on a fraction of those planets, and even if intelligence only evolves on a fraction of the planets with life, there are still likely to be thousands of life-bearing planets in our galaxy. Finding such life will, however, mean finding the planets. Even if interstellar travel was routine, the job would be daunting. It would mean finding one world among thousands, with no evidence of its special status visible at interstellar distances.

INTELLIGENT LIFE IN OUR GALAXY

Intelligent life, if it exists elsewhere, is likely to be much rarer than nonintelligent life. It may, however, prove easier actually to *find*. Our own species beams a steady stream of radio and television signals into space and attaches information-laden metal plates to spacecraft headed out of the solar system. The signals are an accidental by-product of broadcasting; the plates are a conscious attempt at communication. Both announce our existence, our level of technological sophistication, and a tiny bit about our culture. The Search for Extra-Terrestrial Intelligence (SETI) program, begun in 1959, uses large radiotelescopes to listen for evidence that other species might be doing similar things.

It is also possible that a sufficiently intelligent and technologically adept species might find us before we develop the ability to go looking for it. Believers in the extraterrestrial origin of UFOs argue that such encounters have already happened, either in the past or in the present. Most mainstream scientists are skeptical of such beliefs, explaining purported encounters with aliens in more prosaic terms.

EXTRATERRESTRIAL LIFE IN POPULAR CULTURE

Popular culture depicts thousands of human encounters with extraterrestrial life. Entire subgenres of science fiction are devoted to such encounters: "first contact" stories, "alien invasion" stories, "aliens among us" stories, and so on. A detailed discussion of popular culture's treatment of aliens could easily fill a book. Nearly all stories about extraterrestrial life, however, follow three well-established conventions.

First, most stories featuring imagined extraterrestrial life tend to focus on one or, at most, two species from any given world. Gatherings of intelligent aliens from many worlds are common (the barroom scene from *Star Wars* [film 1977] is a classic example), but fully imagined alien ecosystems are not. The reason for this is both obvious and understandable. Ecosystems are extraordinarily complex. Describing one on Earth, the building blocks of which are familiar, is a significant challenge; creating a plausible alien ecosystem from scratch, using very different building blocks, is an even greater challenge. Stories that meet that challenge—Hal Clement's *Mission of Gravity* (1954), for example, or Larry Niven's *The Integral Trees* (1984)—often bring the world to life at the expense of the characters who populate it. Most storytellers, understandably, prefer to spend their energy on the characters, reducing the environment they inhabit to exotic background.

Second, the physical form of extraterrestrial species reflects human attitudes toward species on Earth. The sweet-natured title character of Stephen Spielberg's film *E.T.* (1982) has a head that is large in proportion to its body and eyes that are large in proportion to its head. It has, in other words, the basic morphology of a human infant: the same morphology that makes Mickey Mouse "cute." Alien species that invade or attack the earth often resemble creatures that Western culture deems unpleasant. The Martian invaders in H.G. Wells's novel *The War of the Worlds* (1899) have tentacles, as do their successors in the movie *Independence Day* (1996). The invaders in the TV series *V* (1983–1985), and in Harry Turtledove's *Worldwar* (1994–1997) and *Colonization* (1999–   ) series of novels are reptilian. Those in *Starship Troopers* (novel 1959, film 1997), Anne McCaffrey's *Rowan* novels (1990–   ), and the TV series *Space: Above and Beyond* (1995–1996) are insectoid. Powerful and benevolent aliens, on the other hand, recall angels in their lack of permanent physical bodies. Their evolution "beyond the need for physical form" is also suggestive of ideas about the afterlife. Creators of such aliens, like Arthur C. Clarke in his four "Odyssey" novels (1968–1997) and J. Michael Straczynski in the TV series *Babylon 5* (1994–1999), often make these parallels explicit.

Third, the personalities and thought patterns of intelligent aliens (the vast majority of those portrayed, for obvious dramatic reasons) closely resemble those of humans. Alien invaders of Earth want what human invaders want: territory, resources, slaves, or mates. Alien benefactors of Earth act out of altruism or paternalism or to secure allies in a hostile universe. Humans and aliens routinely discover that despite their physical differences, they share many of the same hopes and fears. "The Devil in the Dark" (1967), one of the most-beloved episodes of the original *Star Trek* TV series, reaches its climax when the alien "monster" is revealed as a mother protecting its children. Barry B. Longyear's short story "Enemy Mine" (1979, film 1985) involves a human and an alien, opponents in a long, bitter war, who become friends after their ships crash on the same planet. Finding common ground between multiple intelligent species was a running theme of the entire *Star Trek* canon, as well as of *Babylon 5*, Keith Laumer's stories of interstellar diplomat Jaime Retief (1963–   ), and Spider Robinson's "Callahan's Bar" stories (1973–   ). The assumption in all such stories is that common ground *does* exist and that intelligent beings of goodwill can find it.

Genuinely *alien* aliens are rare in popular culture, partly because it is difficult to create them, partly because it is difficult for audiences to identify with them. They are rare enough, in fact, that well-drawn examples readily stand out from run-of-the mill aliens. Robert L. Forward's novel *Dragon's Egg* concerns an alien race, the Cheela, that live on the surface of a neutron star under gravity 67 billion times greater than that of Earth. The cheela evolve rapidly, a generation taking only thirty-seven minutes, and contact with a human survey ship triggers their twenty-four-hour rise to civilization. They have, by the end of the book, far surpassed their slow-moving human teachers. "The Emissary" (1993), the pilot episode for *Star Trek: Deep Space Nine* (1993–1999), introduces disembodied aliens who (unlike humans) do not experience time as linear. Past, present, and future events are for them all coexistent. Their conversations with the series' principal human character, Benjamin Sisko (Avery Brooks), are refreshingly alien in their elliptical structure and enigmatic content. Even these stories, however, assume a basic level of congruence between human and alien thoughts.

We know nothing of how extraterrestrial life—if it exists—appears, behaves, or (if intelligent) thinks. Stories about it thus allow for limitless imagination. We tend, nevertheless, to imagine aliens whose appearance reflects our attitudes toward species here on Earth and whose thought and behavior patterns mirror our own. The reason for this is less a failure of imagination than an acknowledgement of dramatic necessity.

Stories about human encounters with alien species are, ultimately, stories about us rather than the aliens. The otherworldly visitors in the TV series *Third Rock from the Sun* (1995–2001), like the unworldly title character of Voltaire's *Candide*, comment on human eccentricities from an outsider's perspective. The implacable invaders of *V* and *Independence Day* allow the human characters to demonstrate their bravery and resourcefulness. The innocent, stranded aliens of films like *Escape from the Planet of the Apes* (1971), *Starman* (1984), and *Brother from Another Planet* (1984) are litmus tests for human society. Good-hearted individuals shelter and aid them, but those in power persecute them; the stories simultaneously reveal the best and worst of human behavior. Stories like these require aliens that are more human than any real alien species is likely to be—aliens that are human enough for human characters to interact with and for human audiences to care about.

Related Entries: Evolution; Evolution, Convergent; Life, Origin of; Mars; Space Travel, Interplanetary; Space Travel, Interstellar; UFOs; Venus

FURTHER READING

Darling, David. *Life Everywhere: The Maverick Science of Astrobiology*. Basic Books, 2001. Argues enthusiastically that life is common; compare Ward.

Davies, P.C.W. *Are We Alone? The Philosophical Implications of the Discovery of*

Extraterrestrial Life. Basic Books, 1995. Brief, wide-ranging overview of the extraterrestrial-life debate.

Levay, Simon, and David Koerner. *Here Be Dragons: The Scientific Quest for Extraterrestrial Life*. Oxford University Press, 2000. Another solid overview, including organisms thriving in extreme environments on Earth.

Pickover, Clifford A. *The Science of Aliens*. Basic Books, 1999. Breezy speculations, grounded in science, on extraterrestrial life.

Ward, Peter Douglas, and Donald Brownlee. *Rare Earth: Why Complex Life Is Uncommon in the Universe*. Copernicus Books, 2000. Argues that the circumstances that make life on Earth possible are complex and therefore unusual; compare Darling.

Life, Origin of

The earth is approximately 4.5 billion years old. The oldest known traces of single-celled life come from rocks 3.6 billion years old, and the first single-celled organisms probably arose even earlier. The conditions under which life arose on Earth are well understood. The cooling of the earth's crust allowed liquid water to collect on its surface, forming oceans. Gasses from the primordial atmosphere—ammonia (NH_3), methane (CH_4), carbon dioxide (CO_2), hydrogen (H_2), and water vapor (H_2O)—dissolved into the primordial ocean, along with sulfur and phosphorus ejected by volcanoes. The primordial oceans thus contained the basic chemical building blocks of life: carbon, hydrogen, oxygen, nitrogen, sulfur, and phosphorus. The oceans were warmed by the sun, stirred by storms and currents, and periodically infused with energy by lightning strikes and solar radiation.

Stanley Miller and Harold Urey showed in a famous 1953 experiment that organic molecules form readily in a laboratory simulation of this "primordial soup." Subsequent experiments have confirmed these results and refined scientists' understanding of the process. How these molecules—amino acids, lipids, DNA-like strands—became organized into single-celled creatures capable of self-reproduction is less clear. The most popular explanation is that hundreds of millions of years of mixing and combining molecules brought together the essential elements of a cell by chance. A variation of this theory holds that the molecules forming the cell's interior formed simultaneously with sheet-like lipids that, when folded around the cells, became the cell walls. Another variation holds that lipids formed first and that some (by chance) became folded around tiny bubbles of "primordial soup," from which other molecules arose.

Not all scientists accept the idea that early life emerged from the random mixing of organic molecules. Some argue that the impact of comets or asteroids into the primordial ocean played a role, catalyzing the formation of organisms or depositing already-formed organisms in the nutrient-rich ocean. Others, advocates of the "intelligent design" theory, believe that the formation of the first cells was the purposeful, creative act of a sentient being. Most scientists argue that the latter theory is outside the realm of science and that the former merely begs the question "How, then, did the organisms from space form?" The standard model of the origins of life, shorn of its biological details, appears frequently in popular culture. Moreover, like science, popular culture usually acknowledges the existence of a first single-celled organism, while admitting ignorance of its exact origins. Jane Robinson's "Origin of Life Drinking Song" works a dozen different scenarios into its rapid-fire verses. Arthur Guiterman's poem "Ode to the Amoeba" (1922) invites readers to honor the very first single-celled animal: "The First Amoeba, strangely splendid/From whom we're all of us descended." Bugs Bunny, asked by a TV interviewer to recount his life "from the start," launches into a wildly dramatic description of primordial Earth. At its climax, he lowers his voice reverently to call attention to "two tiny amoebas . . . *the start of life.*"

The idea of life emerging spontaneously from lifeless material is also well established in popular culture. A song by Chris Weber (lyrics published in *Isaac Asimov's Science Fiction Magazine*, May 1982) urges listeners to "Beware of the sentient chili/That bubbles away on the stove." One installment of Jim Davis's comic strip *Garfield* begins with the cat's owner peering into his refrigerator and declaring that it needs to be cleaned out; Garfield sardonically agrees, observing that "the bologna is grazing on the lettuce." Stories of life created by accident in the laboratory convey a similar message: life *wants* to emerge, and given anything close to the right conditions, it will.

Theories that the emergence of life on Earth was orchestrated rather than spontaneous get significantly more attention in popular culture than in science. The most common variation involves technologically advanced aliens who "seeded" Earth (and, often, other worlds) with primitive life forms. Spider and Jeanne Robinson's "Stardancers" trilogy of novels (1979, 1991, 1995) uses this premise, as does the movie *Mission to Mars* (2000). The creators of *Star Trek: The Next Generation* invoked the theory, in the 1993 episode "The Chase," to explain the extraordinary number of intelligent humanoid species in the *Trek* universe. The idea that divine intervention was involved in the origin of life is also more common in mainstream popular culture than in mainstream science. The psychiatrist-hero of the 1980 film *The Ninth Configuration* delivers a long monologue on the improbability of life arising spontaneously. Dr. Harry Wolper of *Creator* (novel 1980, film 1985) argues implicitly that the power to create life from nonlife lies with God (or people, like himself, with godlike powers).

Related Entries: Evolution; Evolution, Convergent; Life, Extraterrestrial; Religion and Science

FURTHER READING

Ellington, Andrew. "Interim FAQ: The Probability of Abiogenesis." *Talk Origins Archive*. 6 April 1995. Talk Origins. 10 May 2001. <http://www. talkorigins.org/faqs/faq-abiogenesis.html>. A detailed outline of key biochemical steps in the origin of life.

Fry, Iris. *The Emergence of Life on Earth: A Historical and Scientific Overview*. Rutgers University Press, 2000. Brief, wide-ranging introduction to the origin-of-life debate.

Schopf, J. William. *Cradle of Life*. Princeton University Press, 1999. Designed for readers with high school-level biology background.

Lightning

Lightning is a natural form of electrical discharge. It occurs in times when and places where the atmosphere is highly charged with electricity, and it can take many forms: streaks, sheets, balls, or the glow called "St. Elmo's Fire" that sometimes engulfs ships and aircraft. Lightning flashes between charged clouds and the ground roughly 20 million times each year in the forty-eight contiguous states alone. Cloud-to-cloud flashes are five to ten times more common. Cloud-to-ground strikes can reach up to fifteen miles from their points of origin, with enough force to start fires, explode sap-laden trees, or kill large animals (including humans). Lightning kills 100 people and injures another 250 in an average year in the United States. Fires kindled by lightning strikes cause damage, in an average year, that reaches into the billions of dollars.

The relentless mechanical churning of water droplets that takes place in thunderstorms concentrates a positive electrical charge in the upper layers of the cloud and a negative electrical charge in the lower. Cloud-to-cloud lightning strikes occur when oppositely charged parts of two clouds come close enough for the attraction between the two charges to overcome the distance between them. Cloud-to-ground strikes occur when a negatively charged cloud bottom passes close to a positively charged object on the ground. Slender, pointed objects accumulate positive charge at their tips, making them natural targets for lightning strikes. Lone trees, steeples, antennas, and masts are all susceptible. So are standing human beings, especially those holding golf clubs or umbrellas above their heads. The standard advice for surviving lightning if caught in the open—stay low and curl into a ball—is designed to mask the lightning-friendly shape of the human body.

Traditional ideas about the origins of lightning, formed long before anyone thought of it as electricity, focused on its genuine destructive power and its presumed rarity. Lightning symbolized nature's capriciousness and was, in myth, the preferred weapon of sky gods like Odin and Zeus. Lightning's image in popular culture owes more to these ancient traditions than to modern science. The idea that it "never strikes twice in the same place" is wrong but deeply rooted in the prescientific idea that nature is inherently random.

Ancient tradition and modern popular culture acknowledge, as modern science does, the destructive power of lightning. They go farther, however, by giving lightning metaphysical powers as well as physical ones. Fictional lightning often transforms, rather than destroys, the object it strikes. It gives stitched-together creatures the spark of life in countless variations of *Frankenstein*. It brings "Number 5," the robot-hero of the *Short Circuit* movies (1986, 1988), the gift of self-awareness. One lightning bolt transports the hero of L. Sprague de Camp's novel *Lest Darkness Fall* (1941) from the twentieth century to the fifth, and another—channeled from point of impact to a time machine by wires carefully placed beforehand—allows time travelers from 1985 to return "home" from 1955, in the first *Back to the Future* movie (1985).

Neither science nor Judeo-Christian scripture portrays lightning bolts as weapons wielded by angry gods against specific targets. The idea persists in popular culture, however. Characters in Johnny Hart's comic strips *B.C.* and *The Wizard of Id* are periodically struck by enormous lightning bolts as retribution for ill-considered words or actions. An oft-told joke has a golf-playing priest repeatedly shouting "Damn, I missed!" when his putts miss the hole and being cautioned by his partner that such blasphemy will anger God. A bolt of lightning obliterates the pious member of the twosome, and a voice booms, "Damn, I missed!"

Intense lightning storms are a routine part of summertime weather in much of the United States, especially the Southeast and Midwest. In popular culture, however, intense displays of lightning are nearly always meaningful; they show that great and terrible powers are at work. The half-mad Captain Ahab addresses his crew in a scene from Herman Melville's *Moby-Dick* (1851), while St. Elmo's Fire wreaths masts, rigging, and the harpoon in his hand. The heroes of Richard Wagner's operas often appear amid flashes of lightning, and "What's Opera, Doc?" (1957), Chuck Jones's merciless cartoon parody of Wagner's operas, puts Bugs Bunny at the mercy of Earth-splitting lightning bolts controlled by archenemy Elmer Fudd. Lightning storms are so common in tales of horror and the supernatural that they have been genre cliches for decades. Popular music also uses lightning to symbolize intense emotions. It signals the flaring passion of two lovers in the Lou Christie/Twyla Herman rock anthem "Lightin' Strikes" (1965) and the murderous anger of a betrayed wife in the Garth Brooks/Pat Alger ballad "The Thunder Rolls" (1992).

Whatever else it might be in popular culture, lightning is almost never

Lightning as divine justice. The Roman god Jupiter strikes down the corrupt, the self-important, and the foolish in an 1868 lithograph by Frederick Heppenheimer. Courtesy of the Library of Congress.

just electricity arcing from cloud to ground. It is too spectacular, too impressive, and too laden with cultural meaning to be only that.

Related Entries: Electricity; Life, Origin of

FURTHER READING AND SOURCES CONSULTED

"Jumpstart: Understanding Lightning." 2001. ScienceMaster.com. 7 April 2001. <http://www.sciencemaster.com/jump/earth/lightning_bolt.php>. Complete introduction; source of the scientific data in this entry.

Uman, Martin. *All about Lightning*. Dover Books, 1986. Answers to fifty common questions; based on Uman's *Lightning* (1983)

"Weather: Understanding Lightning." 26 March 2001. USAToday.com. 7 April 2001. <http://www.usatoday.com/weather/thunder/wlightning.htm>. Brief, well-illustrated discussions, emphasizing safety tips.

Longevity

Americans can now reasonably expect to live between seventy and eighty years. Headstones in colonial-era graveyards show our ancestors regularly reaching similar ages. The Ninetieth Psalm, written over 2,000 years earlier, gives the same figure: "threescore years and ten," or "four-score" for the strong. Modern medicine has substantially improved *average* life expectancy by reducing deaths from infections and communicable diseases, especially among young children. It has not, however, made the oldest Americans of today a great deal older than the oldest Americans of 200 years ago. More Americans than ever before reach old age, but lives exceeding eighty-five years are still atypical, and lives exceeding 100 are still rare.

The human body, even in the absence of serious injury or acute illness, simply wears out after seventy or eighty years. Bones become brittle, joints degrade, circulation diminishes, eyesight and hearing fail, and brain functions begin to falter. The many who live full and active lives well into their eighties and nineties generally do so in spite of, rather than in the absence of, increasing physical limitations. A healthy lifestyle, practiced over a lifetime, can lessen the impact of those limitations. So can medication, as well as mechanical devices ranging from eyeglasses to pacemakers and artificial joints. Making century-plus lives the norm will require more extensive changes: radical alterations in diet, more extensive drug therapies, and tinkering with the body's biochemistry at the molecular level. Extending life spans well into a second century may require systematically swapping worn-out organs for mechanical or genetically tailored replacements. Molecule-sized "nanomachines," which proponents say will one day be capable of repairing the body cell by cell, are another potential tool for doubling or tripling current life spans. Like organ replacement, they would require biological, medical, and mechanical knowledge well beyond the current state of the art.

Popular culture seldom challenges the idea that "threescore years and

Year of Birth	Estimated Life Expectancy
1900	47.3
1910	50.0
1920	54.1
1930	59.7
1940	62.9
1950	68.2
1960	69.7
1970	70.8
1980	73.7
1990	75.4

Life expectancy of U.S. citizens at birth, 1900–1990. Statistics are for both sexes and all races. Statistics for 1900, 1910, and 1920 are based on average age at death. Data courtesy of the U.S. Bureau of Vital Statistics.

ten" is a normal human lifetime. Centenarians are as comparatively rare in fiction as they are in real human populations. Individuals who live well beyond a century appear solely in stories of the fantastic, and even there they are depicted as exceptions rather than the norm, individuals on whom longevity is bestowed by magic, nature, or science.

The magical sources of long life are diverse. The residents of Shangri-La, the hidden kingdom of James Hilton's novel *Lost Horizon* (1933; films 1937, 1973) owe their longevity to the Himalayan enclave where they live. Drinking from the Holy Grail—the cup used by Jesus at the Last Supper—keeps a medieval knight alive until the 1930s in *Indiana Jones and the Last Crusade* (1989). Extreme longevity can also be imposed on an individual, as a form of cosmic justice. The medieval legend of the "Wandering Jew" tells of a man who, having spurned Christ on the day of His crucifixion, is condemned to walk the earth until Judgement Day. Modern versions of the legend abound: Richard Wagner's opera *The Flying Dutchman* (1843); Barry Sadler's novel *Casca: The Eternal Mercenary* (1979) and its sequels; and an episode of the TV anthology series *Night Gallery* (1970–1972), in which a cowardly survivor of the *Titanic* drifts alone in a lifeboat—destined to be "rescued" only by other doomed ships.

The "natural" mechanisms used to account for extreme longevity in popular culture are as varied as the unabashedly magical ones. At least one late-middle-aged female character in Ben Bova's novels *Moonrise* (1996) and *Moonwar* (1998) slows aging by using nanomachines to purge the plaque from her arteries and the wrinkles from her face. The longevity seekers of Aldous Huxley's novel *After Many a Summer Dies the Swan* are stuck with a less appealing method: eating fish entrails. The characters in Robert Heinlein's novels *Methuselah's Children* (1941) and *Time Enough for Love* (1973) have been selectively bred for centuries-long lives. The heroes of James Gunn's novel *The Immortals* (1962) and the similar TV drama *The Immortal* (1970) owe their longevity to a rare mutation that affects their blood. Flint, the millennia-old citizen of Earth featured in the original *Star Trek* series's "Requiem for Methuselah" (1968), is another lucky mutant. His body is capable of "instant tissue regeneration," which renders him—unlike many long-lived fictional characters—immune to injury as well as disease.

Stories in which great longevity is bestowed through magic nearly always leave the details of the magic obscure. The same is, nearly always, true of stories in which great longevity is a quirk of nature or a product of advanced science. The actions of magic, nature, and science are in fact virtually indistinguishable in such stories: all bestow, by mysterious means, the gift (or curse) of a life span well beyond "threescore years and ten."

Related Entries: Cryonics; Cyborgs; Miniaturization; Mutations; Organ Transplants

FURTHER READING

Bova, Ben. *Immortality: How Science Is Extending Your Life Span and Changing the World*. Avon, 2000. Celebratory popular science treatise on anti-aging research.

Kluger, Jeffrey. "Can We Stay Young?" *Time*, 25 November 1996. 13 December 2000   <http://www.time.com/time/magazine/archive/1996/dom/961125/medicine.can_we_stay_you8.html>. Focuses on the effects of an artificially lowered metabolism.

Slusser, George, Gary Westphal, and Eric S. Rabkin, eds. *Immortal Engines: Life Extension and Immortality in Science Fiction*. University of Georgia Press, 1996. Scholarly articles written from a literary, analytical perspective.

Magnetism

Magnetism, like electricity, is a form of electromagnetic energy. It occurs naturally in both the earth itself and in such minerals as magnetite (also called lodestone). Magnets can also be created artificially, either by bringing the magnet-to-be into repeated contact with an existing magnet or by passing an electric current through it. Magnetism manifests itself as a field that surrounds the magnetized object and has both north and south poles. The unlike poles of two magnetic fields will attract one another, and the like poles will repel one another. One result of this is that small magnets, like compass needles, will align themselves with the magnetic fields created by larger magnets, like the earth. Magnets are essential components of electric motors, electric generators, and the "Second Industrial Revolution," those machines spawned in the late nineteenth century. They are also ubiquitous parts of consumer electronics: telephones, loudspeakers, tape recorders, and computer disk drives.

Magnets are omnipresent in popular culture, because they are vital components of machines that are themselves omnipresent. The magnets in such devices remain invisible, however. Characters in popular culture (like their real-world counterparts) treat the complex machines of everyday life as "black boxes": they know what goes in and what comes out, but not what happens inside. Even when magnet-dependent machines break down, the magnets inside are rarely the source of the problem. The compass, which *does* suffer magnet-related malfunctions, is an exception to this "black box" pattern in the real world, but not in popular culture. Novice compass users learn that they must correct for errors from two sources: regional differences between true north and magnetic north ("declination" or "variation"), and nearby metal objects and electromagnetic fields ("deviation"). The compasses used in popular culture, on the other hand, are apparently immune from both forms of error. Except when a false reading is essential to the story, they give precise and accurate readings, with no need for correction.

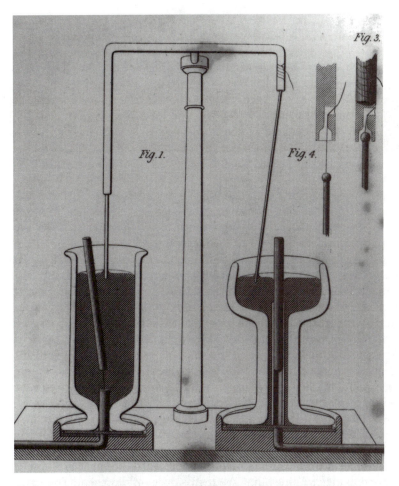

Faraday creates the Second Industrial Revolution. In the 1850s, English physicist Michael Faraday showed, with an experimental apparatus like this one, that the attraction and repulsion between electrically generated magnetic fields could be made to produce rotary motion. His discovery led to the electric motor, one of the key technologies of the second half of the nineteenth century. Courtesy of the Library of Congress.

Magnets rarely appear in the foreground of popular culture. Those that do are often enormous and invariably powerful in relation to their size. Magneto (Ian McKellen), chief villain of the movie *X-Men* (2000), is a mutant who can generate magnetic fields powerful enough to pull down fences and rip heavy doors off their hinges. The magnets that routinely appear in gadget-heavy James Bond movies are slightly more conventional but no less powerful. *You Only Live Twice* (1967) includes a chase scene where Bond is pursued by a large sedan full of Japanese gangsters. The chase ends when Bond's allies use a helicopter, a large electromagnet, and a cable to pluck the gangsters' car from the road and drop it into the nearby sea. Bond himself uses a large magnet in *The Spy Who Loved Me* (1977), snaring a steel-toothed adversary named "Jaws" and depositing him in a convenient tank of sharks. In *Live and Let Die* (1973), Bond is issued a watch containing an electromagnet powerful enough, he is told, to deflect the path of a bullet. He completes his mission without having to test the claim—which is fortunate, since the lead from which most bullets are made is unaffected by magnetic fields.

James Bond movies are essentially live-action cartoons, and *X-Men* is an adaptation of a long-running comic-book series. Neither, therefore, is obliged to observe strictly the laws of physics that govern the real world. Animated cartoons can ignore those laws altogether or rewrite them for comic effect. The magnets shown in such cartoons are, as a result, powerful to the point of comic absurdity.

Bugs Bunny, fighting off a small-scale alien invasion in "Lighter than Hare" (1960), watches a three-robot "demolition squad" drop explosive charges into a ventilation pipe near his home. He produces a large magnet and throws it down the pipe after the explosives. The robots, which have been fleeing the scene at high speed, are pulled backward by the magnet and sucked into the pipe to be blown up by their own explosives. Wile E. Coyote, self-proclaimed "super genius," turns a giant electromagnet against Bugs in "Compressed Hare" (1961). The iron carrot that Bugs was *supposed* to have swallowed comes whistling over the horizon into the coyote's lair, followed by a stream of ever-larger metal objects. Pots and pans give way to refrigerators, bulldozers, ocean liners, and (eventually) satellites pulled out of their orbits high above the earth. Finally, the magnet drags an enormous rocket off its launch pad and into the cave, where it explodes and brings the scene to a close.

Related Entry: Electricity

FURTHER READING

Asimov, Isaac. *Light, Magnetism, and Electricity.* New American Library, 1969. Nontechnical discussion of magnetism, electricity, and their interaction.

Livingston, James D. *Driving Force: The Natural History of Magnets.* Harvard University Press, 1997. Readable, comprehensive overview of the science and technological applications of magnetism.

Macauley, David. *The New Way Things Work.* Houghton Mifflin, 1998. Whimsically illustrated explanations of magnet-using technology.

Mars

Mars, the fourth planet from the sun, is roughly half the size of the earth and twice the size of the moon. The surface of Mars would be a harsh but tolerable environment for properly suited human explorers: Martian gravity is roughly 40 percent of Earth's, the Martian atmosphere (composed mostly of carbon dioxide) is a hundred times thinner than Earth's, and Martian surface temperatures range from −200° F at the poles to 80° F at the equator. What remains of the once-substantial Martian water supply is now locked in polar ice caps and, some scientists believe, in subsurface ice deposits. These deposits may provide a habitat for simple forms of life well below the cold, dry surface sampled by the robot *Viking* (1976) and *Pathfinder* (1997) landers.

Depictions of Mars in popular culture express no such doubts about the presence of life on Mars. Life has been part of the mythology of Mars since the early twentieth century. Astronomer Percival Lowell saw what he thought were artificial canals on the Martian surface and imagined a dying race desperately scavenging water from the polar icecaps. Novelist H.G. Wells unleashed Martian invaders on an unsuspecting earth in *War of the Worlds* (1897). Pulp fiction writer Edgar Rice Burroughs began the swashbuckling adventures of John Carter, "the greatest swordsman of two worlds," in *A Princess of Mars* (1912). Those images—a dying world, a mortal enemy, and a new frontier—have defined Mars in popular culture ever since.

Even before *Mariner* and *Viking* spacecraft returned the first detailed images of the Martian surface, Lowell's image of a water-starved world cast a long shadow. Cdr. Kit Draper, the hero of *Robinson Crusoe on Mars* (1964), is marooned in a place far more desolate than the Daniel Defoe character after whom he is patterned. His "island" has neither plants nor running water, and the "ocean" surrounding it is the vacuum of space. The film was shot in the most desolate of all North American landscapes, Death Valley, California. The utter dryness and desolation of Mars is

Mars seen from orbit. Mars is now a dry world, but surface features like Valles Marineris (roughly parallel to the long edge of the photograph) were shaped in the distant past by running water. Courtesy of the National Space Science Data Center and Dr. Michael C. Malin, Principal Investigator, Mars Observer project.

also central to Theodore Sturgeon's classic short story "The Man Who Lost the Sea" (1959). Its hero, a lone astronaut dying on the sandy surface of Mars amid the wreckage of his spaceship, immerses himself in memories of his first explorations, skin diving in the warm seas of Earth. Only on Mars—a world of beaches without oceans—can he truly appreciate the beauty of his own world.

Mars's proximity to Earth, blood-red color, and association with the Roman god of war make it an ideal launching pad for fictional invasions of our world. The invasions, which began with H.G. Wells's *War of the Worlds* in (1897), have not ended yet. Martian weapons lay waste to London in Wells's novel, to New York in the 1938 radio adaptation by Orson Welles, and to Los Angeles in the 1953 film version. Martian invaders enslave the entire human race in C.M. Kornbluth's short story "The Silly Season" (1950) and seek human mates in the irresistibly titled film *Mars Needs Women* (1968). The invaders adapt to Earth in many ways: remaining cloaked in their machines in *War of the Worlds*, taking possession of human bodies in *Invaders from Mars* (films 1953, 1986), and encasing their bulbous heads in fishbowl-style helmets in the trading-card series (and 1996 film) *Mars Attacks!* Ultimately, however, nearly all are defeated. The people of Earth, with rare exceptions like Kornbluth's "Silly Season," emerge from the experience sadder and wiser, but free.

Stories about the settlement of Mars outnumber even those about Martian invasions. Like stories set on America's Western frontier, they often focus on encounters between pioneers and natives. The Martians of Ben Bova's novels *Mars* (1992) and *Return to Mars* (1999), and of Robert A. Heinlein's story "The Green Hills of Earth" (1947), are dead and gone, leaving only mute buildings behind. The heroes of Heinlein's young-adult novel *The Rolling Stones* (1952) export Martian "flat cats" to miners in the Asteroid Belt, and the hero of his similar *Red Planet* (1949) paves the way for an interspecies treaty of friendship by befriending a beach ball–like Martian named Willis. *The Martian Chronicles* (1950), a series of linked stories by Ray Bradbury, takes a more pessimistic view that echoes the European encounters with the natives of North America. Terrestrial diseases wipe out most of the Martians, and the survivors drift (or are pushed) to the outer edges of the earth colonies. Contact with humans obliterates their culture and characteristic thought patterns so completely that even when humans temporarily abandon Mars, the old Martian ways cannot be revived. Only much later, when a new group of colonists resettles Mars, do humans begin to understand the depth and complexity of the now-extinct Martians.

Related Entries: Life, Extraterrestrial; Moon; UFOs; Venus

FURTHER READING

Arnett, Bill. "The Nine Planets: Mars," <http://seds.lpl.arizona.edu/nineplanets/nineplanets/mars.html>. Images and data from one of the Web's premier science resources.

Raeburn, Paul, and Matt Golombek. *Mars: Uncovering the Secrets of the Red Planet.* National Geographic Society, 1998. Synthesizes data gathered through 1997; superbly illustrated.

Sheehan, William. *The Planet Mars: A History of Observation and Discovery.* University of Arizona Press, 1996. Focuses on the cultural dimensions of astronomical observations.

Matter Transmission

Two of the greatest technological breakthroughs of the industrial age involve transportation. The first—begun by the steam engine, extended by the internal combustion engine, and completed by the jet engine—made it possible to move people and objects faster than an animal could walk. The second—begun by the telegraph and extended in turn by the telephone, television, and fax machine—made it possible to transmit complex messages over great distances at the speed of light. Matter transmission, if developed, would be the spiritual (if not technological) descendent of those breakthroughs. It would allow us to move physical objects (including people) with the same ease that we now move information: virtually instantaneously, over great distances, without damage or corruption.

Matter transmission lies so far beyond our current understanding of science and engineering that we don't know whether it is possible, much less how to make it happen. Logic suggests two different approaches to the problem. The first involves distorting space in such a way that the transmitted object could pass from point A to point B without actually crossing the intervening distance. The second involves breaking down the transmitted object into component atoms at point A and transmitting the atoms to point B, where they are reassembled into their original form. Both methods would require energy in mind-boggling quantities. Both depend on untested assumptions about nature: the first, that humans can warp space at will; the second, that the position, motion, and chemical state of every atom in a transported object can be recorded (and then recreated) simultaneously.

Characters in popular culture who travel by matter transmitter do not worry about these problems. They don't wonder whether their matter transmitter is going to work any more than we wonder if our car is going to move when we press the accelerator. Like cars, however, matter transmitters can be dangerous if they malfunction, and popular culture tends to focus on them at those moments.

The 1958 film *The Fly* remains the single most famous story about matter transmission gone wrong. Its scientist-hero uses himself as the first human test subject for the matter transmitter he built, unaware that a housefly has entered the chamber with him. The experiment scrambles man and fly, leaving the scientist (with a fly's head and arm) searching for the fly (with a man's head and arm). The 1986 remake of *The Fly* replays the basic story but scrambles man and fly at the genetic level, turning the scientist into an oozing, decaying human-fly hybrid. Both films underscore the horrible results of the accident by using the scientist's fiancée as an eyewitness. Both imply that (the 1959 sequel *Return of the Fly* notwithstanding) the technology is a dead end.

The problems created by matter transmitters in science fiction stories like Isaac Asimov's "It's Such a Beautiful Day" (1954) and Larry Niven's "Flash Crowd" (1973) are social rather than personal. Asimov's story is set in a world where near-universal ownership of matter transmitters has eliminated the need to go outside. The hero is a boy who, forced to walk to school when his family's transmitter breaks down, rediscovers the joys of the outdoors. Niven's story involves public matter transmitters installed for the benefit of pedestrians on the corners of city streets. It explores the unexpected (and unintended) aid they give to criminals, for whom a clean getaway is now as close as the nearest street-corner transmitter. Both stories deal, like *The Fly*, with unexpected problems created by matter transmitters. They treat the problems more optimistically, however—as fresh challenges rather than fatal flaws.

Star Trek (five TV series, nine movies, and scores of novels to date) occupies a curious middle ground between these two positions. Its matter transmitters, the now-famous "transporters," suffer spectacular malfunctions on a regular basis. Hapless users have been split, merged, trapped in dematerialized form, and beamed into alternate universes. At least ten major characters have, at one time or another, suffered near-fatal transporter accidents. Despite this hair-raising track record, nobody in the *Star Trek* universe ever appears worried about being transported. Nobody pauses, looks thoughtful, or makes a joking remark as they step on the platform. The only two characters to break this pattern do so for reasons not related to the transporter itself. Dr. "Bones" McCoy, from the original series (1966–1969), is a crusty techno-skeptic out of step with the twenty-third century. Lt. Reginald Barclay, from *Star Trek: The Next Generation* (1987–1994), is patently neurotic.

Star Trek characters' attitude toward transporters is, in general terms, similar to that of many users of risky technologies. Americans, as a society, think in similar terms about cars, which kill and injure tens of thousands every year: the risks are great, but the benefits make the risk worth running.

Related Entries: Space Travel, Interstellar; Time Travel

FURTHER READING

Krauss, Lawrence. *The Physics of* Star Trek. Basic Books, 1995. Chapter 5, "Atoms or Bits," discusses the famous "transporter."

Niven, Larry. "The Theory and Practice of Teleportation." In *All the Myriad Ways*. Ballantine, 1971. A nonfiction article by the most scientifically careful writer of matter-transmission stories.

Woolf, Ian. "Teleportation: Or the Penetration of the Wave Function into a Classically Forbidden Region." 25 August 1997. 20 August 2001. <http://www-staff.socs.uts.edu.au/~iwoolf/txt/teleport.txt>. Despite the forbidding subtitle, an accessible discussion of matter transmission in science fiction.

Meteorites

First, a few definitions: a *meteoroid* is a chunk of rock or metal moving through interplanetary space; a *meteor* is the streak of light visible in the night sky when a meteoroid enters Earth's atmosphere and, heated by friction, begins to vaporize. Most of the thousands of meteoroids that strike Earth's atmosphere are small enough to disintegrate before reaching the ground. A *meteorite* is one of the few hundred each year that survive to strike the ground. Most are comparatively small and leave few visible traces of their impacts. Giant meteorites, capable of forming craters several miles in diameter, are rare. Three are estimated to strike the earth every million years, and only one of those typically strikes land.

Meteorites are made of rock formed at the same time as the earth. They retain their original chemical composition and physical structures—long since erased from the rocks of the still-active earth—and so provide a valuable window on the early geological history of the solar system. Meteorites have also in recent decades been linked to the mass extinctions that punctuate the history of life on Earth. The father-and-son team of Luis and Walter Alvarez proposed in a 1980 paper that an extraterrestrial impact triggered the demise of the dinosaurs and many other species about 65 million years ago. Geological surveys of Mexico's Yucatan Peninsula have since located a possible "smoking gun" near the town of Chicxulub, an impact crater of the right age and size to account for the mass extinction. The crater, partially covered by the Caribbean Sea, is roughly 100 miles in diameter.

Popular culture's interest in meteorites revolves, broadly speaking, around the same features: their extraterrestrial origins and their ability to cause destruction. In fact, popular culture often combines the two attributes by creating meteors whose literally "unearthly" powers spread destruction.

The extraterrestrial-invasion movie *Day of the Triffids* (1963) begins with a meteor shower that blinds most of the human race and then, for

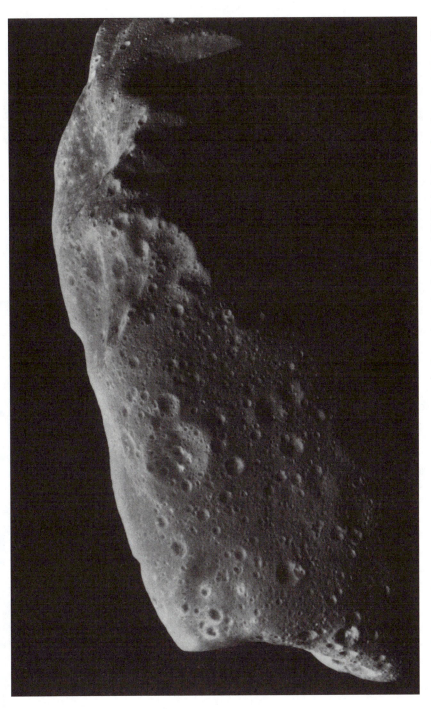

Asteroid 243 Ida. Meteorites are fragments of comets, asteroids, and similar bodies. 243 Ida is believed to have a composition similar to many meteorites recovered on Earth. Courtesy of the National Space Science Data Center and Dr. Robert W. Carlson, Principal Investigator, Galileo project.

good measure, carpets Earth with the alien creatures of the title. The alien force that takes possession of a bulldozer in Theodore Sturgeon's short story "Killdozer" (1944, TV film 1974) comes to Earth on a meteorite. It lies dormant on an isolated island for thousands of years, until an unsuspecting construction crew strikes the meteorite with a bulldozer blade. "The Lonesome Death of Jordy Verrill," one of the installments in the horror-anthology film *Creepshow* (1982), also uses a meteorite as a Trojan horse. This time the stone from space is hollow, filled with a noxious green goo that quickly turns the meteorite's unlucky discoverer into a semi-alien monster. Inner-city teacher Jefferson Reed (Robert Townsend) fares better in the film *Meteor Man* (1993). When a meteorite hits him in the stomach, he gains modest superhuman powers—the ability to fly four feet off the ground, for example.

The films *Meteor* (1979) and *Armageddon* (1998) and Arthur C. Clarke's novel *The Hammer of God* (1993) take a different approach. In each, the earth is threatened by an impending meteorite impact, and the heroes must find a way to avert disaster. The threat in each case comes not from mysterious powers or alien life forms but from the implacable laws of physics. If they cannot destroy or divert the approaching lump of rock, its mass and speed will be sufficient to kill millions at the moment of impact and millions more in the ecological catastrophe that follows. The two films, choosing spectacle over science, end with the approaching rock shattered by nuclear explosives but ignore the shotgun blast of impacts that would result as the fragments hit the earth. Clarke's novel plays the game more honestly; its characters nudge the meteoroid onto a new trajectory that will take it past Earth. A closely related set of stories posits that meteorites may someday be used as weapons. Rebellious lunar colonists fling rocks at the earth in Robert Heinlein's novel *The Moon Is a Harsh Mistress* (1966), and the alien Centauri subject the home world of their rivals, the Narn, to a devastating bombardment in the second-season finale of the TV series *Babylon 5* (1992–1998).

Meteorites' appeal to modern-day storytellers is easy to understand. The thunderbolt-throwing gods of the ancient world are dead, but meteorites make a plausible substitute. Like the ancient gods, they descend from the sky without warning, bestowing gifts or curses on the mortals who encounter them, or spreading destruction in the blink of an eye.

Related Entries: Comets; Inertia

FURTHER READING

Alvarez, Walter. *T. Rex and the Crater of Doom*. Vintage Books, 1998. Overview of the meteorite-extinction theory by one of its authors.

Norton, O. Richard. *Rocks from Space; Meteorites and Meteorite Hunters*. Mountain

Press. 2nd ed. 1998. Lavishly illustrated introduction to both meteorites and comets.

Plait, Phil. "Phil Plait's Bad Astronomy." <http://www.badastronomy.com>. An astronomer's critique of astronomical errors in movies and TV shows, including *Armageddon* and *Meteor*.

Mind Control

Societies teach their newest members, whether children or immigrants, how to behave in socially acceptable ways. Organizations educate their newest recruits in the proper ways of dressing, speaking, and acting while on the job. Members of established households explain the "house rules" to new residents and expect them to adjust their behavior accordingly. These activities form one end of a spectrum that, at its other end, includes torture, brainwashing, and other forms of psychological abuse. All are forms of mind control. All use basically the same psychological principles: breaking down old behavior patterns and establishing new ones, then systematically discouraging the reversions to the old and rewarding displays of the new. All have the same basic goal: to "reprogram" the individual's responses in ways that benefit the group overseeing the process.

Both teachers and torturers practice a form of mind control over their charges. The differences between them, like the differences between surgeons and murderers, are primarily social: one accepts ethical limits, acts on a consenting subject, and pursues a socially acceptable goal; the other does not. The boundary between laudable "socialization" and criminal "brainwashing" is drawn differently in every society. The side of the line on which a specific case of mind control lies is, especially in democratic societies, often sharply contested.

Popular culture generally depicts the results rather than the mechanics of mind control. Because of this, and because it tends to focus on the extreme ends of the spectrum, socialization and brainwashing appear in popular culture as completely distinct processes. Their common features are evident only in those relatively rare works that explicitly ask where the boundary lies.

Stories of individuals learning to be members of a sympathetic group (like a sports team or military unit) portray mind control as a brightly glowing good. The Teacher—often a classroom teacher, but sometimes a coach, priest, boss, or commanding officer—imposes a harsh regimen on the Student, who initially rebels but eventually submits and becomes a better person as a result. The Student, at the end of the story, thanks the Teacher for the gift thus bestowed: the chance to be a good and useful person. The magnitude of the gift justifies, in retrospect, behavior by the Teacher that the Student once saw as irrational and abusive. Luke Skywalker grumbles throughout *Star Wars* (1977) and *The Empire Strikes Back* (1980) about the seemingly pointless exercises his Jedi Knight mentors demand of him. Only in the conclusion of the trilogy, *Return of the Jedi* (1983), does he realize their purpose. Learning "the way of the Force" has, by then, saved not only his life but also his soul.

Stories of individuals "brainwashed" into serving the needs of a despised group (like a cult or totalitarian government) portray mind control as a pitch-black evil. The Master—spymaster, mad scientist, cult leader, or other fanatic—relentlessly crushes the Victim's individuality and strips away the Victim's moral compass. The Victim is no longer fully human but merely a mindless pawn in the Master's evil enterprises. Sgt. Raymond Shaw, the central character in *The Manchurian Candidate* (novel 1959, film 1962), does not even realize that he *is* brainwashed, much less that the process made him a Chinese-controlled assassin-in-waiting. Capt. Jean-Luc Picard, hero of television's *Star Trek: The Next Generation* (1987–1994), unknowingly kills thousands of his comrades in battle during his temporary "assimilation" by the alien Borg collective.

The idea that "socialization" and "brainwashing" amount to the same thing runs like a subversive current through popular culture. Mark Twain's *Huckleberry Finn* (1884) ends with its hero planning to "light out for the territories" in order to escape the civilizing influence of his aunt. Popular songs like Malvina Reynolds's "Little Boxes" (1962) and Supertramp's "The Logical Song" (1979), along with much of Pink Floyd's landmark album *The Wall* (1979), decry what they see as society's systematic campaign against individuality. Anthony Burgess's novel *A Clockwork Orange* (1962) and Stanley Kubrick's film *Full Metal Jacket* (1987) make similar points but cut deeper. Alex, the young man who is the nominal "hero" of Burgess's novel, is a rapist and killer who is reprogrammed into a model citizen after his capture and imprisonment by the state. The heroes of Kubrick's film are, in contrast, ordinary young men reprogrammed into ruthless killers by the state in preparation for service in the Vietnam War. The reprogramming, undertaken in each case for socially acceptable reasons, is brutal and literally dehumanizing. Alex loses his free will, the marine recruits their senses of right and wrong. Both works suggest that government bent on controlling citizens' minds—even for the best of reasons—has moved well into, perhaps through, the moral gray area that separates classrooms from torture chambers.

Related Entries: Psychic Powers; Superhumans

FURTHER READING

Cialdini, Robert. *The Psychology of Persuasion*. Quill, 1993. Persuasion from the perspective of social psychology; discusses both its use to promote social harmony and its misuse by cults.

Pratkanis, Robert, and Elliot Aronson. *The Age of Propaganda: The Everyday Use and Abuse of Persuasion*. Freeman, 2001. Discusses the techniques and applications of mind control on a society-wide scale.

Winn, Denise. *The Manipulated Mind: Brainwashing, Conditioning, and Indoctrination*. Malor Books, 2000. Nonspecialist's introduction to the darker side of mind control.

Miniaturization

The steady miniaturization of electronic devices was among the most striking technological trends of the last fifty years. Computers, once capable of filling entire rooms, now fit comfortably on desktops and in briefcases. Five-function calculators, once the size of a paperback book, can now be as small as a credit card. Mobile phones, once carried in small suitcases, now slip easily into coat pockets. The transition from bulky vacuum tubes to densely packed microchips made this revolution possible, by allowing ever-more-capable machines to be fitted into ever-smaller packages. Market demand drove the trend. Consumers embraced the miniaturized devices' greater portability and ease of use, while appliance and automobile manufacturers seized the opportunity to upgrade their products with compact electronic "brains" and controls.

Electronic devices, whose principal moving parts are electrons, are comparatively easy to miniaturize. Miniaturizing mechanical devices is more complex, since the structural properties of levers, gears, and beams change as they shrink. "Nanomachines," named for the fact that their dimensions are measured in billionths of a meter, have until recently been little more than microscopic sculptures, static and incapable of useful work. Biochemical motors fueled by the same reactions that power living cells are now being developed, however, and nanomachines driven by such motors may have practical applications in medicine and manufacturing within a few decades (Wilson).

The popular culture of the late twentieth century reflected the public's steadily growing enthusiasm for miniaturized electronics—particularly laptop computers and cellular phones. Laptops and cell phones used as plot devices became a cinematic cliché in the 1990s, appearing not only in thrillers like *Disclosure* (1994) and *Independence Day* (1996) but also in romantic comedies like *One Fine Day* (1996) and *You've Got Mail* (1998). Nanomachines, though far less common, are prominent in such science fiction as Ben Bova's novel *Moonrise* (1996) and TV's *Star Trek: Voyager* (1994–2001).

Popular culture's interest in miniaturization centers, however, not on the gradual shrinking of machines but on the instantaneous shrinking of people. Stories about miniature heroes have a long history, from Jonathan Swift's *Gulliver's Travels* (1726) through E.B. White's *Stuart Little* (1945) to Walt Disney Studios' film *A Bug's Life* (1998). Stories in which the heroes are shrunk from normal size have added dramatic value; the heroes, like the audience, find that the once-familiar everyday world has become a dangerous, alien landscape. The 1957 movie *The Incredible Shrinking Man*, for example, turns the steadily diminishing hero loose in his own home. The size of a mouse, he must flee from his own cat. Later, smaller still, he defends himself against a hungry spider with a sewing needle that is, for him, the size of a lance. *Honey, I Shrunk the Kids* (1989) uses computer-assisted special effects to present more spectacular situations. The four children of the title, smaller than ants, find that their suburban yard has become a jungle; when the sprinkler comes on, it becomes a swamp. They flee from ants, ride a bumblebee, and narrowly escape from the tornado-like vortex created by a lawnmower.

Miniaturized humans also act as tour guides to various parts of the natural world. The title character of George Gamow's book *Mr. Tompkins inside the Atom* (1939) is a mild-mannered British civil servant, temporarily shrunken to subatomic size for didactic purposes. The movie *Fantastic Voyage* (1966) is nominally an adventure story about miniaturized doctors entering the bloodstream of a dying scientist in order to save his life. It works far better, however, as a from-the-inside tour of the human body. Joanna Cole and Bruce Degen's "Magic School Bus" series of children's science books sometimes begin with Ms. Frizzle, the world's most remarkable teacher, shrinking herself, her class, and the bus in preparation for a field trip. Their miniaturized excursions have taken them inside beehives, ant colonies, and (in order to learn about germs) a classmate named Ralphie.

Creating ant-sized humans is impossible for the same reason that creating human-sized ants is impossible. Biological systems are not readily scalable; anatomical structures and physiological processes that work well for an organism of a given size would break down if the organism were substantially larger or smaller. Stories about miniaturized people tacitly acknowledge this by not even *attempting* to explain the transformation scientifically. The radioactive mist of *The Incredible Shrinking Man* and the complex machines of *Fantastic Voyage* and *Honey, I Shrunk the Kids* are nothing more than magic wands that, when waved, transport the characters into the fantastic realm of the very small.

Related Entries: Insects, Giant; Robots

FURTHER READING AND SOURCES CONSULTED

Drummer, G.W.A. *Electronic Inventions and Discoveries*. 4th rev. ed. Institute of Physics Publications, 1997. History of electronic components and devices.

Rotman, David A. "Will the Real Nanotech Please Stand Up?" *Technology Review*. March/April 1999, 44–56. A broad survey attempting to separate, too conservatively for some commentators, reality from "hype" and "science fiction."

Wilson, Jim. "Shrinking Micromachines." *Popular Mechanics*, November 1997. 14 December 2000. <http://popularmechanics.com/popmech/sci/9711STROP.html>. Introduction to nanomachines, with excellent color illustrations.

Miracle Drugs

Modern drug therapy began with "salvarsan," developed by Dr. Paul Ehrlich as a treatment for syphilis. Salvarsan, introduced in 1911, was the first drug to attack the root cause of the disease it treated. Its astonishing effectiveness earned it the nickname "Dr. Ehrlich's Magic Bullet"—a reflection of the public's perception of it as a modern miracle. Two mid-twentieth-century pharmaceutical triumphs solidified the public's belief in "miracle drugs." The first was penicillin and, by extension, the many other antibiotics that followed it. Antibiotics reduced a bacterial infection from a life-threatening crisis to a brief unpleasantness, ending a long era in which any injury or surgery that broke the skin carried a substantial risk of death from secondary infection. The second was the Salk polio vaccine, and by extension the vaccines that followed it for measles, mumps, rubella, and other childhood diseases. These vaccines, administered through massive vaccination campaigns tied to public school attendance, had spectacular results. Diseases that had once killed infants and young children by the thousands all but vanished from the industrialized world within a generation.

Penicillin and the polio vaccine raised public expectations of what drugs in general could do. Their rapid, highly publicized successes and their lack of obvious, significant drawbacks fostered a belief in drug therapies as a kind of modern-day magic, capable of vanquishing any disease, no matter how terrible, in a single stroke. The rapid introduction of new drugs and equally rapid control of old diseases created another expectation, that science could develop drugs to prevent or cure *any* disease. Popular culture strongly reflects both expectations.

Medical dramas routinely use doctors as heroes, and why-is-this-patient-sick puzzles as the crux of their plots. The climax of such plots comes when the hero solves the puzzle and (as a result) realizes how to cure the patient. The cure and the recovery that follows are brief epilogues to the main plot, less dramatically interesting, because they in-

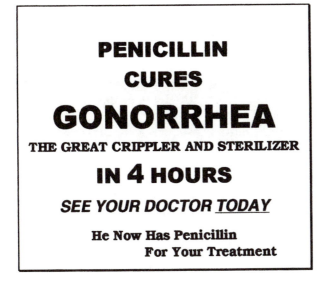

PENICILLIN
CURES
GONORRHEA
THE GREAT CRIPPLER AND STERILIZER
IN 4 HOURS
SEE YOUR DOCTOR TODAY
**He Now Has Penicillin
For Your Treatment**

Advertising penicillin. Reproduction of a poster promi-
nently displayed on city street corners during World
War II. Original image courtesy of the U.S. Public
Health Service.

volve the hero only peripherally or not at all. Both realistic medical stories like the TV series *ER* (1994–) and fanciful ones like the movie
Outbreak (1995) routinely cut, therefore, from the administration of a
drug to evidence that the patient is recovering. Adventure stories where
the dramatic focus is on getting the drug to the patient (as in Tom God-
win's famous 1942 science fiction story "The Cold Equations") work in
similar ways. Once the obstacles are overcome and the patients *have* the
drug, their quick recovery is assumed or shown in a brief closing scene.
The cumulative effect of both kinds of stories is to emphasize the speed
and effectiveness with which the drugs work, enhancing their "miraculous" image.

Drug manufacturers—now able, because of deregulation, to advertise
prescription medications in the mass media—play on the same kinds of
expectation. Their advertisements typically show beautiful people enjoy-
ing life to the fullest amid beautiful settings, implying that this is possible
because a new drug has freed them from the effects of an unpleasant or
debilitating medical condition. The seniors can play with their grandchildren because their arthritis is held at bay; the young woman can walk
through fields of flowers because her allergies are controlled; the middle-
aged man can enjoy his wife's company because his sexual dysfunction
has been cured. Viewers are urged to "ask their doctor" whether the

drug is right for treating *their* condition. If it is, the ads imply, their lives too can be miraculously improved.

Dozens of diseases remain unconquered or uncontrolled; popular culture and public expectation hold, however, that they soon will be, if only enough time and resources are devoted to research. Dozens of charities raise money for research on specific diseases, from Parkinson's disease and muscular dystrophy to AIDS, by appealing to donors' faith that a cure can be found if we only look hard enough. Drug companies, lobbying against price-control legislation in the late 1990s, argued that reducing their profits would shrink their research budgets and slow development of new "miracle drugs."

The public's faith that a pharmaceutical "magic bullet" exists (or can be found) for every disease is also evident in two common expressions. "How dare we spend money on [an allegedly frivolous government program]," runs the first, "when we still don't have a cure for [a well-known disease]?" The disease invoked is typically cancer or, less often, AIDS. "We can [achieve a great technological breakthrough]," laments the second expression, "but we can't cure the common cold." Both statements imply that our lack of will is responsible for the lack of a cure. The evolution of drug-resistant bacteria and the vaccine-defeating variability of the cold virus suggest another possibility: that our microscopic enemies may have us outmaneuvered.

Related Entry: Epidemics

FURTHER READING

Drews, Jurgen. *In Quest of Tomorrow's Medicines.* Trans. David Kumer. Springer-Verlag, 1999. Drug research at major pharmaceutical companies.

Levy, Stuart B. *The Antibiotic Paradox: How Miracle Drugs Are Destroying the Miracle.* Perseus Press, 1992. The evolution of antibiotic-resistant bacteria.

Porter, Roy. *The Greatest Benefit of Mankind: A Medical History of Humanity.* Norton, 1998. The definitive history of medicine's role in Western society.

Moon

The moon is the brightest object in the night sky, roughly a quarter of the size of the earth and slightly less than a quarter-million miles away. These are, for everyone but planetary geologists, probably the three most interesting things about it. The moon's mass gives it a gravity one-sixth that of Earth—low enough to make space launches easy, high enough for humans to work in comfortably. Its distance from Earth puts it nearly a hundred times closer than the two nearest planets, Venus and Mars. Even with chemical rockets, it takes only a few days to reach the moon; the same trip to Mars, made under the best possible conditions, would take at least eight months. The moon has always been part of humankind's mental universe. It is, by solar system standards, an easy place to get to and an attractive place to be. The first attempts to live and work on another world are likely to be made there.

The moon has two different faces in popular culture. The first, painted in the bright silver-white of moonlight, is that of a magnificent, unreachable object that looks down on humankind from above. The second, painted in the dull, dusty gray of moon rocks, is that of a real, tangible place that humans could actually visit. The first image is older than recorded history; the second is little older than the beginning of modern rocketry in the 1930s.

The first image of the moon stresses the aspects of it most clearly visible from the earth: its brilliant white reflected light and its twenty-seven-day cycle of phases. Here, the moon is nearly always divine and nearly always female. The Greeks linked it to Artemis, the virgin goddess of the hunt, whom the Romans called Diana, associating its white light with her sexual purity. Pagan religions throughout Europe regarded the moon as a powerful agent of transformation, noting that its cycles of phases marked the stages of the farmer's year and paralleled women's menstrual cycles. The Judeo-Christian creation story pairs the moon and sun in Genesis 1:16: "God made two great lights: the greater light to rule

Exploring the moon. Dwarfed by the landscape, like the human figures in nineteenth-century paintings of the American West, an Apollo
17 astronaut collects geological samples in the Taurus-Littrow valley.
Courtesy of the National Space Science Data Center and Dr. Frederick
J. Doyle, Principal Investigator, Apollo 17.

the day, and the lesser light to rule the night." The pairing is for many commentators a symbolic statement about gender roles: woman does not create light but reflects the light she receives from man. Alfred Tennyson took this view in his poem "Locksley Hall" (1842): "Woman is the lesser man and all thy passions, match'd with mine/Are as moonlight unto sunlight, and as water unto wine."

The moon's associations with women and purity remained strong in the nineteenth and twentieth centuries but gradually lost their religious overtones. The moon, still female, is today more a watchful mother than a distant goddess. It looks down on the sleeping bunny in Margaret Wise Brown's classic children's book *Goodnight Moon* (1947) and, voiced by Lynne Thigpen, listens to Bear recount the day's adventures in each episode of the Disney Channel TV series *Bear in the Big Blue House* (1998–   ). Lovers meeting by moonlight—the oldest cliche in popular music—also feel the moon's watchful maternal eye. No matter how intense their passion, no matter what they might do elsewhere, they share only hugs and tastefully restrained kisses while the moon is looking down on them.

Popular culture's second image of the moon portrays it as a new frontier—a place to be explored, settled, and domesticated. This image is intimately tied to memories of another, now-closed frontier, the American West. Stories set on and around the moon routinely recycle characters, events, and situations originally found in Western locales, adapting technological and environmental details. The most inventive of these stories recast the familiar elements in novel ways. The most derivative simply swap spaceships for stagecoaches and stun guns for six-shooters.

Some stories about the lunar frontier consciously echo history. D.D. Harriman, the businessman who pioneers commercial space travel in Robert Heinlein's "Future History" stories, shares both the name and entrepreneurial spirit of Union Pacific Railroad tycoon E.H. Harriman. Other stories retell Western legends. Frontier veterans dealing with clueless greenhorns, stranded explorers desperately low on supplies, and the just-in-time arrival of the cavalry have all been translated to the lunar frontier. A third group of stories, like the made-for-TV movie *Salvage*, uses the moon to dramatize Western values. When its inventor-hero uses a homebuilt spaceship to retrieve valuable hardware abandoned on the moon by NASA, he demonstrates the superiority of practical know-how to "book learning" and of plucky individuals to sluggish government agencies. All three types of stories carry a distinctly American theme: there will always be new frontiers to challenge would-be pioneers.

Related Entries: Comets; Mars; Venus

FURTHER READING AND SOURCES CONSULTED

Arnett, Bill. "The Moon." *The Nine Planets.* 13 June 2001. <http://seds.lpl. arizona.edu/nineplanets/nineplanets/luna.html>. Brief but comprehensive description of the moon, with links.

Long, Kim. *The Moon Book: Fascinating Facts about the Magnificent, Mysterious Moon.* Rev. and expanded ed. Johnson Books, 1998. Brief, reader-friendly introduction, emphasizing astronomy and folklore.

Spudis, Paul D. *The Once and Future Moon.* Smithsonian, 1998. The best serious, book-length introduction for nonspecialists.

Mutations

Mutations are changes in the structure of organism's genes or in their arrangement on the chromosomes. Genetic mutations—those that alter a gene's structure—can produce new variants (alleles) of the gene that produce new traits in the organism. Chromosomal mutations—those that alter a gene's position on the chromosome—can cause bits of hereditary information to be eliminated, duplicated, or scrambled. This too can produce new traits in the organism. Spontaneous mutations occur slowly but steadily in nature. The mutation rate increases if the organism is exposed to agents called mutagens—X-rays, gamma rays, ultraviolet radiation, and certain chemicals.

Some mutations have no discernable effect on the organism's viability. Some mutations are so disruptive that the organism is nonviable or sterile. Those that fall between the two extremes provide the raw material on which natural selection, the mechanism that drives evolution, operates. Mutations that handicap an organism in its struggle for survival tend to disappear over the course of many generations. Mutations that benefit an organism in its struggle for survival tend not only to persist but to spread through the population. Favorable mutations, while rare, thus tend to be preserved; unfavorable ones, while more common, tend to fade away.

The changes wrought by most real-world mutations are limited in scale. Large ones are usually fatal or, at best, prone to produce sterility or crippling illness. Mutations that affect genes outside the reproductive cells have even narrower effects, often cancer or other diseases. Popular culture takes a more expansive view. Positive mutations in humans are giant evolutionary leaps forward that give once-ordinary individuals superhuman powers. Negative mutations in humans produce physically degenerate subhumans. Mutations in animals make innocuous creatures threatening and threatening creatures terrifying. The line between mutants and nonmutants is always sharply drawn in popular culture. The conflicts that drive stories about mutants often take place across that line.

Popular culture treats positive mutations in humans as valuable gifts. Spiderman, the Incredible Hulk, and the Fantastic Four (among costumed superheroes) owe their powers to mutations created by radiation exposure. The Teenage Mutant Ninja Turtles, their teacher, and the henchmen of their chief enemy all became human-animal hybrids after contact with "mutagenic ooze." The X-Men and *their* principal adversaries were born as mutants, as were the heroes of the "Wild Cards" series of science-fiction stories edited by George R.R. Martin. Characters with psychic powers are often described as mutants; so too are those with extraordinary intelligence or self-healing ability. All of these mutation-derived talents are "improvements" on ordinary humans—part of "natural" evolutionary progress and thus acceptable. Even when, as in the *X-Men* saga, mutants act in the service of evil, their mutant status amplifies their evil deeds rather than causing them. Humans with positive mutations may be feared and persecuted by "normal" humans, but audiences are invited to sympathize with the mutants. Stephen King is especially adept at creating such sympathy. Even in *Carrie* (1974), where the heroine commits mass murder in the climax, it is hard *not* to cheer for the persecuted mutant.

Negative mutations in humans typically result from a great catastrophe, and they often leave their victims grotesquely scarred. Despite this, the victims get little or no sympathy in popular culture. Their mutations also scar their souls, driving them to monstrous acts that threaten the stories' nonmutant heroes. The mutants in Richard Matheson's novel *I Am Legend* (1956), filmed as *The Omega Man* (1971), are vampires. Those in George Romero's "Living Dead" films are blank-eyed zombies hungry for human flesh. The mutants of the film *Battle for the Planet of the Apes* (1972), horribly scarred by radiation, live underground and harbor an irrational hatred of the civilized ape society on the surface. Mutants, the stories imply, act as they do *because* they are mutants. The heroes are thus justified in taking violent, even genocidal, action against them.

Mutations in animals are, like negative mutations in humans, usually tied to a disruption in the normal workings of nature. Nuclear weapons tests produce giant ants in the movie *Them!* (1954); an earthquake inexplicably unleashes fire-starting insects in Thomas Gage's *The Hephaestus Plague* (1973, filmed in 1975 as *Bug*); and pollution gives rise to giant, deformed forest-dwelling beasts in the film *The Prophecy* (1979). The mutants themselves are treated in similar terms—as parts of nature but disruptions of its normal and proper workings. Their size, behavior, and origins mark them (like humans with negative mutations) as "unnatural," which makes the heroes' efforts to contain or destroy them laudable.

Ever since *Frankenstein* (1819), popular culture has been deeply concerned with the line between the "natural" and the "unnatural." The differing attitudes it displays toward mutants represent one example of its efforts to maintain that boundary.

Related Entries: Evolution; Genes; Superhumans

FURTHER READING

Gonick, Larry, and Mark Wheelis. *The Cartoon Guide to Genetics*. Rev. ed. Harper Perennial, 1991. Informative introduction to genetics, including mutations.

"Just What Is a Genetic Disorder?" Genetic Science Learning Center. University of Utah. 15 May 2001. <http://gslc.genetics.utah.edu/disorders/definition/index.html>. Carefully written and well-illustrated overview of how mutations cause disease, birth defects, and other medical problems.

Schwartz, Jeffrey H. *Sudden Origins: Fossils, Genes, and the Emergence of Species*. John Wiley, 2000. Discussion, with extensive historical context, of the idea that new species arise suddenly and discontinuously, with mutations in key genes playing a central role.

Newton, Isaac

Isaac Newton was born in England in 1642, a century after the death of Copernicus, a Polish astronomer who introduced the modern, sun-centered model of the solar system, and months after the death of Galileo, the Italian astronomer and physicist who popularized it. Newton became the climactic figure in the Scientific Revolution—a radical transformation of Western ideas about nature that Copernicus began and Galileo extended. His three laws of motion became part of the foundation of classical physics. His theory of universal gravitation established that falling bodies near the surface of the earth are governed by the same force that holds the planets in their orbits. His experimental work in optics helped to create the modern understanding of light and color, including the idea that a beam of white light can be divided into beams of colored light. His invention of the calculus (in parallel with German mathematician Gottfried Leibniz) gave scientists an essential tool for analyzing nature. On a more abstract level, the heavily mathematical nature of his work reinforced one of the Scientific Revolution's basic methodological principles—that, in Galileo's words, "the book of nature is written in the language of mathematics."

Newton's scientific reputation rests on two books. The first, entitled *Mathematical Principles of Natural Philosophy* (1687), integrated his own work on physics with that of his predecessors and contemporaries. Better known as the *Principia* (a short form of its Latin title), it is the literary capstone of the Scientific Revolution. The second, titled *Opticks* (1701), summarized his work on light and color. Written in English rather than Latin and aimed at a broader audience, it is part of an experiment-oriented tradition of physics that began with Galileo and flourished in the eighteenth century. Newton's intellectual output was significantly larger and more varied than these two works suggest, however. In addition to purely mathematical papers, he wrote extensively on biblical interpretation and on alchemy. He was also active in the politics of both

AXIOMATA SIVE LEGES MOTUS

Lex. I.

Corpus omne perseverare in statu suo quiescendi vel movendi uniformiter in directum, nisi quatenus a viribus impressis cogitur statum illum mutare.

PRojectilia perseverant in motibus suis nisi quatenus a resistentia aeris retardantur & vi gravitatis impelluntur deorsum. Trochus, cujus partes cohaerendo perpetuo retrahunt sese a motibus rectilineis, non cessat rotari nisi quatenus ab aere retardatur. Majora autem Planetarum & Cometarum corpora motus suos & progressivos & circulares in spatiis minus resistentibus factos conservant diutius.

Lex. II.

Mutationem motus proportionalem esse vi motrici impressae, & fieri secundum lineam rectam qua vis illa imprimitur.

Si vis aliqua motum quemvis generet, dupla duplum, tripla triplum generabit, sive simul & semel, sive gradatim & successive impressa fuerit. Et hic motus quoniam in eandem semper plagam cum vi generatrice determinatur, si corpus antea movebatur, motui ejus vel conspiranti additur, vel contrario subducitur, vel obliquo oblique adjicitur, & cum eo secundum utriusq; determinationem componitur.

Lex. III.

Lex. III.

Actioni contrariam semper & aequalem esse reactionem: sive corporum duorum actiones in se mutuo semper esse aequales & in partes contrarias dirigi.

Quicquid premit vel trahit alterum, tantundem ab eo premitur vel trahitur. Siquis lapidem digito premit, premitur & hujus digitus a lapide. Si equus lapidem funi alligatum trahit, retrahetur etiam & equus aequaliter in lapidem: nam funis utrinq; distentus eodem relaxandi se conatu urgebit Equum versus lapidem, ac lapidem versus equum, tantumq; impediet progressum unius quantum promovet progressum alterius. Si corpus aliquod in corpus aliud impingens, motum ejus vi sua quomodocunq; mutaverit, idem quoque vicissim in motu proprio eandem mutationem in partem contrariam vi alterius (ob aequalitatem pressionis mutuae) subibit. His actionibus aequales fiunt mutationes non velocitatum sed motuum, (scilicet in corporibus non aliunde impeditis:) Mutationes enim velocitatum, in contrarias itidem partes factae, quia motus aequaliter mutantur, sunt corporibus reciproce proportionales.

Corol. I.

Corpus viribus conjunctis diagonalem parallelogrammi eodem tempore describere, quo latera separatis.

Si corpus dato tempore, visola M, ferretur ab A ad B, & vi sola N, ab A ad C, compleatur parallelogrammum ABDC, & vi utraq; feratur id eodem tempore ab A ad D. Nam quoniam vis N agit secundum lineam AC ipsi BD parallelam, haec vis nihil mutabit velocitatem accedendi ad lineam illam BD a vi altera genitam. Accedet igitur corpus eodem tempore ad lineam BD sive vis N imprimatur, sive non, atq; adeo in fine illius temporis reperietur alicubi in linea illa

Newton's *Principia* (1687). Newton wrote in the style of the ancients, laying out intricate mathematical proofs in formal Latin, but his ideas (on these pages, his three laws of motion) completed the seventeenth-century demolition of Aristotle's view of the universe. Courtesy of the Library of Congress.

the scientific community and the nation, leading the Royal Society, serving in Parliament, and acting as director of the Royal Mint.

The Newton of popular culture is a brilliant but eccentric loner. He is also much narrower than the Newton of history, with no interests except his physics and mathematics. Darwin, as portrayed in popular culture, has his family; Einstein has his political activism; Galileo has his long battle with the Catholic Church. Newton has the study and the laboratory, but nothing outside them. He is purely a scientist and, at that, a scientist in the style of the twentieth or twenty-first century rather than his own seventeenth. Popular culture routinely ignores Newton's contributions to alchemy and biblical scholarship—legitimate parts of science in his day, but not ours. On the rare occasions when it does take note of them, it treats them as proof of eccentricity, not brilliance.

Eighteenth-century observers routinely described Newton as godlike in his insight. A famous couplet by English poet Alexander Pope proclaimed:

Nature, and nature's laws, lay hid in night.
God said: "Let Newton be!" and all was light.

William Blake, an early member of the Romantic movement, rejected Newton's vision of a machinelike universe but acknowledged the power of the mind that had created it. The central figure in his painting "Newton" (1795–1805) bears a strong resemblance to Greek and Roman sculptures of Apollo, and the gray-bearded mathematician-God in his "Ancient of Days" (1794) is also inspired by Newton. The eighteenth-century hero of Jean Lee Latham's young-adult novel *Carry On, Mr. Bowditch* (1955) studies a Latin edition of the New Testament in order to learn enough of the language to read the *Principia*. The word of God becomes for him a stepping-stone to the more complex word of Newton.

Images of Newton in twentieth-century popular culture acknowledge his brilliance but highlight his eccentricity. Harpo Marx's portrays him in *The Story of Mankind* (1957) as a wide-eyed, clownish innocent who derives enlightenment (specifically, the idea of universal gravitation) from an unlikely source: a falling apple. The holographic Newton played by John Neville in "Descent, Part 1," a 1993 episode of *Star Trek: The Next Generation*, is stiff and formal, like a wax-museum statue. Marx's Newton physically distances himself from others by leaving the university for the countryside; Neville's Newton distances himself socially with his prickly personality. When time travelers Peabody and Sherman visit Newton in a segment of TV's *The Adventures of Rocky and Bulwinkle* (1959–1964), they find him so distracted that they must maneuver him into his legendary encounter with the falling apple.

Popular culture's image of Newton fits the role he is assigned in Western culture. Because his work is part of the foundation of modern science, he is seen, accurately or not, as the first modern scientist. It is no surprise, therefore, that he embodies popular culture's most enduring stereotype of scientists—intellectually brilliant but personally eccentric and socially isolated.

Related Entries: Acceleration; Action and Reaction, Law of; Galileo; Gravity; Inertia

FURTHER READING AND SOURCES CONSULTED

Cohen, I. Bernard. *The Birth of a New Physics*. Rev. ed. Norton, 1991. One of many excellent histories of the Scientific Revolution; sets Newton's work in a broader context.

Cohen, I. Bernard, and Richard S. Westfall, eds. *Newton: Texts, Backgrounds, Commentaries*. [A Norton Critical Edition] Norton, 1994. Excerpts from Newton's work, with historical context.

Lucking, David. "Science and Literature." 28 January 2001. 5 May 2001. <http://www.lucking.net/course1.htm>. Easy access to images of the Blake paintings discussed above.

Westfall, Richard S. *The Life of Isaac Newton*. Cambridge University Press, 1994. A condensed, nonspecialist's version of *Never at Rest* (1983), Westfall's definitive biography of Newton.

Organ Transplants

Organ transplants are an established, though far from routine, part of modern medicine. Surgeons can remove such organs as the liver, kidneys, corneas, and heart from the bodies of the recently dead and use them to replace the damaged or diseased organs of still-living patients. The medical barriers to successful organ transplants remain substantial. Organs are highly perishable once removed from the body, requiring precise coordination of two surgical procedures. The surgery itself is lengthy, difficult, and resource intensive. Finally, even if surgery is successful, the recipient's immune system may interpret the new organ as a foreign body and attack it. Screening of organs and recipients for compatibility reduces the chances of rejection but limits the number of organs suitable for any given recipient. Suppressing a recipient's immune system with drugs also reduces the chances of rejection, but it increases the vulnerability of the recipient's already weakened body to infection. Genetically altered pigs may be a promising source of donor organs, but the technology remains experimental ("Designer Donors").

Social factors, such as a perennial shortage of donors and disagreement over how to allocate scarce organs, also complicate transplants. These complications are intensified by the emotions that organ transplants arouse in potential donors and recipients, as well as their families. Organ transplant surgery treads near, and sometimes on, Western beliefs about the sanctity of the body and taboos against its mutilation. The stories that popular culture tells about it—whether realistic, fanciful, or dark— derive their dramatic power from those intense emotions.

Realistic treatments of organ transplants are common on TV medical dramas such *St. Elsewhere* (1982–1988) and *Chicago Hope* (1994–1999). Replacing the old organ with the new is, in most cases, background: "just" one more miracle for the doctor-heroes to perform. The main action of the story involves human interactions outside the operating room, where would-be recipients wait on borrowed time while the families of would-be donors must make agonizing choices about their just-dead loved ones. Even transplant stories at the outer limits of realism tend to focus on the social dimensions of the process.

Fanciful treatments of organ transplants generally dispense with the operation itself early and quickly. Their focus is often a bond that the shared organ somehow creates between donor and recipient. The donor's presence is sometimes physical, as in films like *The Incredible 2-Headed Transplant* (1971) or *The Thing with Two Heads* (1972) but more often spiritual. The hero of the movie *Return to Me* (2000) loses his beloved wife in a car wreck, only to meet (by chance) the young woman into whom his wife's heart was transplanted. Their burgeoning romance is, as both title and plot imply, shaped by the mystical influence of the shared heart. *Heart Condition* (1990) uses a similar device: a bigoted white police officer is shadowed by the amiable ghost of the dead black lawyer whose heart was transplanted into his body. Countless movie adaptations of *Frankenstein*, of course, attribute the monster's violent behavior to the insertion of a murderous criminal's brain into the monster's empty skull.

Dark stories about organ transplants play on the anxieties audiences already feel about the process by giving it horrific new dimensions. Their settings and incidental details are realistic, firmly grounding them in the real world, but terrible events unfold beneath the reassuringly familiar surface. The 1971 film *The Resurrection of Zachary Wheeler* centers on a secret clinic in New Mexico that clones its patients and uses the clones (conveniently grown without brains) as a living, breathing sources of spare parts. Robin Cook's medical thriller *Coma* (novel 1977, film 1978) takes place in a hospital where selected surgical patients are given poisoned anaesthetic and used as sources of transplantable organs for the black market. An apparently immortal urban legend tells of unwary travelers who, after drinking with a stranger in a hotel bar, awaken in an ice-filled hotel bathtub holding a cellular phone and a note urging them to call 911 before they bleed to death. The victim finds, according to the legend, that one of his kidneys is missing—removed by the stranger and his accomplices for sale on the black market.

The kidney-theft legend is wildly implausible: why would thieves ruthless enough to steal a kidney from a living victim go to such great lengths to preserve him and encourage him to call for help? It persists nonetheless, because of its capacity to make audiences shudder. That power is rooted in the same values and taboos that shape public attitudes toward organ transplants and lends emotional power to other, less lurid stories about them.

Related Entries: Cloning; Cyborgs; Longevity

FURTHER READING AND SOURCES CONSULTED

"Designer Donors." *New Scientist.* 25 March 2000. 11 December 2000. <http://www.newscientist.com/hottopics/cloning/cloning.jsp?id=22310400>. The first successful cloning of pigs and its implications.

Gutkind, Lee. *Many Sleepless Nights: The World of Organ Transplantation.* University of Pittsburgh Press, 1990. Doctors' and patients' perspectives, emphasizing emotional issues.

Mikkelson, Barbara. "You've Got to Be Kidneying." *Urban Legends Reference Pages.* June 2000. San Fernando Valley Folklore Society. 11 December 2000. <http://www.snopes.com/horrors/robbery/kidney.htm>. Discusses and debunks the kidney-theft legend, with references.

The Transplant Web. 11 December 2000. <http://www.transplantweb.org>. A comprehensive online resource for organ-transplant issues.

Prehistoric Humans

When used by scientists, "prehistoric humans" is a very broad category. Popular culture uses it more narrowly, as shorthand for the people who lived in Europe during the Paleolithic, or Old Stone Age. The term "Paleolithic" was coined in 1865, to cover the long first act of European prehistory—the thousands of years between the first appearance of humans and the invention of agriculture. The scientists who first used the term gave it connotations of "backwardness" and "primitiveness" that it has never entirely lost. As early as the 1880s, however, archaeological discoveries began to show that the people of Paleolithic Europe were remarkably sophisticated.

The chipped-stone tools that gave the Paleolithic its name have finely worked edges and a variety of shapes that tailor them to specific tasks. They were, in many cases, mass-produced by skilled craftsmen and distributed among the members of a community. The tools also grew more sophisticated over time; broken cobbles and roughly chipped axes gave way to slender blades and fluted spear points. Paleolithic humans also worked in plant and animal products like wood, bone, horn, and skin. Finally and most strikingly, they created art: carvings, body decorations, and the famous cave-wall paintings of Altamira and Lascaux.

Paleolithic societies did not settle in a single location, as their farm-tending successors did, but neither did they wander aimlessly across the land. Many moved with the seasons through a cycle of established camps, returning year after year to places that provided food in abundance. Some groups, at least, carried their sick, infirm, and injured members with them; skeletons reveal individuals who had lived for years after bone-breaking injuries that must have crippled them. Some of the dead, at least, were buried with elaborate decorations that suggest caring and sorrow on the part of the living—and perhaps belief in an afterlife.

Popular culture's image of generic "cavemen" paints prehistoric humans in a less than flattering light. It reflects all the connotations of

Paleolithic tools. Originally little more than crude modifications of naturally broken cobbles, stone tools became both varied and sophisticated by the end of the Paleolithic Era. From Henry Fairfield Osborn, *Men of the Old Stone Age* (Scribners, 1915), 339.

backwardness and primitiveness attached to the Paleolithic in the 1860s, and it makes the caveman a symbol of what modern humans have escaped by becoming "civilized." The caveman looks like a brute and acts like one as well; he "courts" a would-be mate by knocking her senseless with a club and dragging her home by the hair. The prehistoric humans that appear in popular culture as individuals, with names and personalities, are a different matter. They are depicted far more sympathetically, embodying the personality traits that we most value in ourselves and our fellow "civilized" humans. More often than not, beneath their fur clothes and stone tools, they *are* us.

Popular culture's portrayal of prehistoric humans as modern humans in fur clothes is easiest to see when used for comic effect. The four main characters in *The Flintstones* (1960–1970) are clearly meant to be working-class Americans of the 1950s and 60s. Fred and Barney gripe about work, pursue easy money, and relax at bowling matches and lodge meetings. The wives, Wilma and Betty, tend their small tract houses but much prefer shopping. The caveman characters in newspaper comic strips like Johnny Hart's *B.C.* (1958–) and V.T. Hamlin's *Alley Oop* (1933–) are also recognizably modern "types" who behave in familiar ways. Mel Brooks's film *History of the World—Part 1* (1981) shows a caveman labeled "the first artist" finishing a cave-wall painting. The punch line comes when another caveman, labeled "the first art critic," gives the painting a bad review—by urinating on it.

The cavemen that appear in "realistic" depictions of prehistoric life are equally, but less visibly, modern in their attitudes and behavior. The family dynamics of the children's TV series *Korg: 70,000 B.C.* (1974–1975) are little different from those in other family dramas. The singer in Steely Dan's song "The Caves of Altamira" (released on the 1976 album *The Royal Scam*) meditates on cave-wall paintings and realizes that he and

Paleolithic cave painting. Belying popular images of the bestial "caveman,"
Paleolithic-era Europeans created detailed wall paintings of the animals they
hunted. From Henry Fairfield Osborn, *Men of the Old Stone Age* (Scribners,
1915), 368.

the prehistoric artist share the same intense need to create. Ayla, the
heroine of Jean M. Auel's novel *Clan of the Cave Bear* (1980) and its three
sequels (1982, 1985, 1990), displays an Edison-like ingenuity. Separated
from her own tribe, she uses its knowledge and her own considerable
ingenuity to transform the lives of one adopted clan after another. Her
saga is, in a sense, the humans-master-nature story of the Industrial Age,
played out in prehistoric Europe.

Ideas and behavior, as scientists are quick to point out, do not fossilize.
The little we know about how Paleolithic humans lived comes from the
material remains that their lives left behind. We know little about how
they interacted with one another and less about what they thought. Popular culture fills that vacuum with the idea, comforting though hard to
prove, that people of all eras share a common humanity.

Related Entries: Evolution, Human; Prehistoric Time

FURTHER READING

Bibby, Geoffrey. *The Testimony of the Spade*. Collins, 1956. Brilliantly written history of ideas *about* European prehistory.

Gamble, Clive. *The Paleolithic Societies of Europe*. Cambridge University Press, 1999. A comprehensive treatment of Paleolithic lifestyles.

Schick, Kathy, and Nicholas Toth. *Making Silent Stones Speak*. Simon and Schuster, 1993. Later chapters discuss the form and use of Paleolithic stone tools.

Prehistoric Time

Earth, according to scientists' best estimates, is about 4.5 billion years old, and it has probably sustained multicelled life for at least 3.5 billion of those years. The Cambrian Explosion, the flowering of complex organisms that laid the foundations of the living world we know, took place 625 million years ago. Plants took to the land by 400 million years ago, and animals by 345 million years ago. *Tyrannosaurus rex*, arguably the single best-known prehistoric animal, flourished at the close of the Cretaceous period, which ended about 65 million years ago. Humans, geologically speaking, are newcomers to the planet. The first members of our genus, *Homo*, appeared only 2.5 million years ago, the first members of our species, *Homo sapiens*, around 400,000 years ago.

These numbers provide the framework on which scientists build their understanding of the past. They define the tempo of astronomical, geological, and biological change. They are also, for most nonscientists, all but meaningless. "Deep Time," as John McPhee dubbed it in his book *Basin and Range* (1981), is so far removed from our everyday experiences that it is literally incomprehensible. A hundred years is an unusually long human lifetime. Recorded history, the equivalent of fifty such lifetimes lived end to end, taxes our ability to grasp long spans of time. Human prehistory, 500 times longer than the recorded history that came after, exceeds our ability—and the history of life on Earth is 1,000 times longer than *that*. We have no more conception of a million years than we do of a million miles or a million dollars.

The difficulty of comprehending or depicting huge spans of time is evident in popular culture's portraits of the distant past. They often telescope time, merging widely separated eras into a single, vaguely defined "long ago." Generic pictures of "The Age of the Dinosaurs" often show brontosauruses contemplating flowering plants that had yet to evolve, and observed by sail-backed dimetrodons that flourished 50 million years earlier. Generic images of the "Age of Mammals" often show

Date	Event
January 1	Formation of the universe
September 9	Formation of our solar system
September 14	Formation of the earth
September 25	Origin of life on earth
December 17	Multicellular life becomes common
December 19	First vertebrates
December 21	First land animals
December 26–28	Dinosaurs flourish
December 31, 8 P.M.	Human, chimpanzee lineages divide
December 31, 8:45 P.M.	Oldest known human ancestor
December 31, 10:45 P.M.	First members of genus *Homo*
December 31, 11:59 P.M.	First *Homo sapiens* identical to us
December 31, 11:59:20 P.M.	First farming communities
December 31, 11:59:50 P.M.	First writing: recorded history begins

The "cosmic calendar." Carl Sagan's famous illustration of the magnitude of prehistoric time: when the fifteen-billion-year history of the universe is compressed into one calendar year, the 5,000 years of recorded human history become the last ten seconds of New Year's Eve.

mammoths and saber-toothed cats alongside the tiny ancestral horse *eohippus*, which lived tens of millions of years earlier.

The film *One Million Years B.C.* (1939, 1966) shows battles between its spear-wielding heroes and dinosaurs that would have been dead for well over 100 million years by the date specified in the title. The toy company Aurora, which manufactured plastic model kits of dinosaurs in the mid-1960s, included two small caveman figures with its brontosaurus. More placid dinosaurs form part of the sparse background of Johnny Hart's comic strip *B.C.* (1958–) and act as transportation for the cave-dwelling characters in *Alley Oop*, created in 1933 by V.T. Hamlin. The animated TV series *The Flintstones* (1960–1970) made dinosaurs staples of the prehistoric world its human characters inhabited. They are the cranes in the quarry where Fred works, the trucks on the streets of the city, and—like the irrepressible Dino—household pets.

A film that, with a straight face, showed Julius Caesar waltzing with Queen Elizabeth I in the ballroom of the *Titanic* would seem patently absurd. The prehistoric equivalent, humans living with dinosaurs, seems vaguely plausible, because of the lengths of time involved and the difficulty of grasping them. Indeed, polls consistently show that 30 to 40 percent of Americans believe that humans and dinosaurs *did* coexist.

Popular culture's mixing of creatures that actually lived millions of years apart is less a product of ignorance than of artistry. Scenes of tiny humans interacting with huge dinosaurs are full of both dramatic and comic potential. The filmmakers who remade *One Million Years B.C.* in

1966 exploited the former. They realized that, aside from actress Raquel Welch's fur bikini, the duels between humans and dinosaurs were the most striking parts of their picture. Gary Larson, the scientifically literate cartoonist who drew *The Far Side*, often took advantage of the latter. One frequently clipped *Far Side* panel shows a caveman, equipped with flip chart and pointer, briefing his hunting partners on the hazards of the stegosaurus. "This," he says, pointing to the dinosaur's spiked tail, "is called the 'Thagomizer,' after the late Thag Simmons."

Even images of the past meant to be educational and scientifically accurate sometimes distort chronology for artistic effect. Murals of the history of life routinely give the most recent parts of Earth history more space, and the earliest parts less space, than their relative lengths in years would indicate. Doing so allows them to feature prominently familiar, visually interesting species like mammoths, saber-toothed cats, and early humans . . . and to minimize the less-interesting-looking algae, plankton, and jellyfish that dominated the early history of life.

Related Entries: Dinosaurs; Evolution; Prehistoric Humans

FURTHER READING AND SOURCES CONSULTED

Albritton, Claude C. *The Abyss of Time*. Freeman Cooper, 1981. Definitive history of the idea of prehistoric time.

Cattermole, Peter John. *Building Planet Earth: Five Billion Years of Earth History*. Cambridge University Press, 2000. Overview of geological changes on the grand scale.

McPhee, John. *Annals of a Former World*. Farrar, Straus, and Giroux, 2000. Omnibus edition of McPhee's lyrical essays on the geological history of North America, including *Basin and Range*.

Psychic Powers

"Psychic powers" is a blanket term for a wide range of extraordinary mental abilities. It encompasses telepathy (communicating by thought alone), telekinesis ("pushing" objects without touching them), extrasensory perception ("seeing" objects obscured by barriers), precognition ("knowing of" events before they happen), as well as more exotic powers. "Scientific" studies of psychic powers began only in the mid-nineteenth century. Results from them have been greeted, ever since, with enthusiasm on the part of believers and incredulity on the part of skeptics. The intellectual boundary between skeptics and believers corresponds closely with the social boundary between mainstream science and what practitioners call "anomalous science." Arguments across the boundary have both intellectual and social dimensions.

Believers argue that the existence of psychic powers has been confirmed repeatedly by anecdotal evidence, individual demonstrations, and systematic experiments. The first category includes self-reported "premonitions" that prove to be accurate and the apparent ability of some twins to communicate without words. The second includes a wide range of public demonstrations by self-proclaimed "psychics" who appear to read minds, bend spoons, predict the future, identify hidden objects, and perform similar feats. The third includes systematic experiments with multiple, randomly chosen subjects and multiple iterations. The cumulative weight of this evidence, believers argue, is decisive. To them, mainstream scientists' refusal to accept it is proof that their minds are closed to observations that threaten established ideas and their monopoly on truth.

Skeptics argue that no anatomical or physiological basis for psychic powers has ever been found and that the existence of such powers is inconsistent with established natural laws. Neither argument disproves the existence of psychic powers, they argue, but both raise the standard of evidence that must be met in order to prove them. Believers, they contend, have failed to meet even the basic standards of scientific proof—much less the elevated ones. Anecdotal evidence is typically reported after the fact, vaguely documented, and subject to unconscious bias—forgetting "premonitions" that did not come true, for example. Public demonstrations can be (and have frequently been) faked by performers familiar with the techniques of stage magic. Experiments on psychic powers are routinely performed without controls that are considered routine in mainstream science: careful monitoring of subjects, randomization, double-blind testing, and so on. Leading skeptic James Randi offered $10,000 in 1964 to anyone who could demonstrate the existence of psychic powers or other paranormal phenomena under scientifically controlled conditions. The money, since raised to a million dollars, has yet to be claimed. Skeptics argue that believers' charges against mainstream science are specious, efforts to discredit standards they know they cannot meet.

Popular culture sidesteps most of the debate; it simply assumes that psychic powers exist. Characters with such powers are standard features in science fiction, fantasy, and horror stories, and the stories often focus on the effects that their powers have on their lives. Classic examples include the telepathic hero of Alfred Bester's *The Demolished Man* (novel 1953), the pyrokinetic heroine of Stephen King's *Firestarter* (novel 1980, film 1984), and the multitalented protagonists of Anne McCaffrey's "Federated Teleport and Telepath" novels. TV series from *Bewitched* (1964–1972) through *Babylon 5* (1993–1998) have featured characters with psychic abilities, and many others have made occasional use of them. Psychic powers are often spread through entire fictional families by heredity or conferred on rooms full of people by government experiments. They are also, often, latent—capable of being awakened by any number of accidents. Max Fielder (Chevy Chase) becomes telekinetic when splashed with toxic waste in *Modern Problems* (film 1981), and Nick Marshall escapes from near electrocution with the ability to read women's thoughts in *What Women Want* (film 2000). The fictional world is, evidently, awash in psychic powers.

So, according to popular culture, is the real world. Advertisements and "infomercials" touting the services of self-proclaimed psychics are common on late-night television. Filmed performances by psychics are common on the smaller networks, as are credulous documentaries devoted to anecdotal evidence of psychic phenomena. Nostradamus, a sixteenth-century physician and astrologer who claimed the ability to foretell the future, is the subject of a small but thriving publishing industry, more than 100 books currently in print claim to interpret his often cryptic visions. Tabloid newspapers, notably the *National Enquirer*, routinely begin the new year by offering their psychics' predictions for the coming twelve months (evaluations of the predictions for the year just ending are conspicuously absent). Finally, fortune-tellers continue to do a brisk business, both on fair midways and in storefronts, as they have for centuries.

It is ultimately impossible to say whether the market for such things is driven by primarily by genuine belief or a search for entertainment. The two may not even be separable. Part of the appeal of psychic phenomena and other aspects of "fringe science" has always been that believing in them is more fun than *not* believing.

Related Entries: Mind Control; UFOs

FURTHER READING

Hines, Terrence. *Pseudoscience and the Paranormal: A Critical Examination of the Evidence.* Prometheus Press, 1988. Debunking analysis of psychic research and other anomalous science.

Hyman, Ray. *The Elusive Quarry: A Scientific Appraisal of Psychical Research.* Prometheus Books, 1989. Critical evaluations of experiments with psychic phenomena.

Schwecker, Michael. *Best Evidence.* iUniverse.com, 2001. Compilation of the most compelling evidence for psychic powers and other anomalous science.

Race

Races are varieties of humankind. They are products of a time when the ancestors of modern humans lived in small, isolated groups that seldom encountered or mated with each other. The most visible differences between races are physical—variations in skin pigmentation, facial features, and hair. On the basis of such features, scientists have divided humankind into as few as three major races or as many as twelve. Most Americans recognize at least four: white, black, Asian, and Indian. American attitudes toward race changed profoundly over the course of the twentieth century, and science both drove and reflected those changes. All scientific investigations are intertwined with the culture of the time and place where they are done, investigations of humankind more than others, and investigations of race and gender most of all.

A sea change in American attitudes toward race began in the early 1940s and extended through the 1970s. The black, Latino, and American Indian civil rights movements were significant, but not sole, causes of it. The defeat of overtly racist regimes in imperial Japan and Nazi Germany, the dissolution of Europe's Asian and African empires, and the dismantling of America's own "empire" in the Philippines also cast doubt on old, easy assumptions about race. So too did the highly publicized achievements of supposedly "inferior" races on the battlefields and playing fields of the 1940s and 1950s. The resulting shift in American ideas about race was slow and frequently erratic, but it was real. The images of race reflected in the dual mirrors of science and popular culture at the beginning of the century were fundamentally different from those reflected at the end.

Early in the twentieth century, many Americans believed that racial differences ran far deeper than the hair and skin: that each race had characteristic abilities, levels of intelligence, and personality traits. Many Americans believed that race defined what individuals could achieve. The cultural differences between nations and the relative social positions of racial groups within a given nation thus reflected biological differences. Some believed that if nonwhite Americans lacked social prestige, economic prosperity, and political power, it was only a reflection of the substantial "natural" differences between the races. Mainstream American popular culture reflected this view of race in two interrelated ways.

First, popular culture portrayed nonwhites almost exclusively through stereotypes, as standard sets of traits and quirks rather than as individuals. Blacks were portrayed as dull-witted servants and lackeys who remained freakishly content and carefree even when living threadbare lives, but babbled or fled in terror at the slightest sign of danger. American Indians were seen as bloodthirsty savages who whooped with joy as they burnt and massacred their way across the frontier. Asians were stereotyped as fussy and excitable servants or, in stories like Sax Rohmer's tales of Dr. Fu Manchu, power-mad tyrants scheming to destroy Western (that is, white) civilization. The prevailing scientific view of race sanctioned this stereotyping. If both physical appearance and character were rooted in biology, then all Indians would be warlike, just as all leopards would have spots.

Second, popular culture defined nonwhite characters by their lack of the admirable traits that whites possessed. Whites were portrayed as rational and calm, blacks as emotion-driven and panicky. Whites were seen as peaceful, Indians as warlike. Whites played fair, Asians smiled politely but broke the rules when it suited their purposes. This us-versus-them approach to race relations, which emphasized and amplified racial differences, also drew strength from the prevailing scientific view of race.

Nearly all American scientists who investigated race in the nineteenth century held similar views. They believed that the physical differences between races reflected substantial differences in character and ability, and that the structure of American society reflected those differences. Many continued to hold such views well into the 1920s, and a few publicly defended them as late as the late 1940s. These beliefs about race shaped the results of their investigations—results that, in turn, placed the full authority of science behind American society's existing racial prejudices. As late as the 1920s, for example, scientists who studied human evolution routinely described members of non-white races as "less evolved" and "more animal-like" than whites. These characterizations reflected, and reinforced, the white Americans' widespread belief that blacks, especially, were motivated by emotion and desire rather than reason. Racial prejudice had a particularly strong effect on scientific studies of intelligence. Nineteenth-century scientists who tried to gather data on intelligence by measuring skull volume frequently distorted that data, consciously or unconsciously, in ways that reflected contemporary racial stereotypes.

Critiques of these views began to appear in the early twentieth cen-

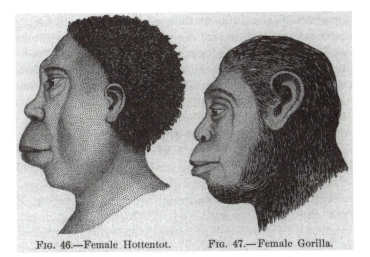

Fig. 46.—Female Hottentot.   Fig. 47.—Female Gorilla.

Scientific racism in the nineteenth century. Scientists who saw
significant biological differences between races often argued
that dark-skinned humans were physically more primitive—
more "apelike"—than light-skinned ones. Illustrations of the
idea, like this one from Alexander Winchell's *Preadamites*, 2nd
ed. (S.C. Griggs, 1880), often distorted both ape and human features to increase their similarity.

tury. Franz Boas, a leading anthropologist, sharply criticized the idea
that race determined culture, as did his colleague Ruth Benedict and her
protegé Margaret Mead. Edna Ferber, who in 1926 brought complex
story lines to the American musical theater with her play *Showboat*, used
it to attack simplistic beliefs about race. Julie, one of the supporting characters, looks and "passes for" white in the post–Civil War South. Because
one of her eight great-grandparents was black, however, the law defines
her as black and forbids her to marry Jim, the white man she loves.
Ferber, through the story of Julie and Jim, attacks the belief (still widespread in 1926) that "blackness" could—like a disease—be conveyed
from one generation to the next even by a "single drop of blood." Isolated views like these set the stage for the much broader shift in attitudes
that began after World War II.

New understandings of human heredity, the discovery of bias in many
nineteenth century studies of race and intelligence, and the realization
that the Nazis had used scientific "proof" of supposed racial inferiority
as a pretext for genocide led scientists to reconsider the biological significance of race in the late 1940s. The new scientific consensus that
emerged in the 1950s assigned only limited importance to racial differences. It proposed that the most significant differences—susceptibility to

certain hereditary diseases, for example—are invisible, and the most visible ones—variations in facial features, hair, and skin pigmentation—are insignificant. This consensus is now so firmly established that attempts to assert a correlation between race and intelligence have met with fierce scientific criticism. Richard J. Herrnstein and Charles Murray's 1995 book *The Bell Curve*, for example, was widely denounced for arguing that differences between blacks' and whites' scores on intelligence tests have a biological basis. Many scientists now argue that the concept of "race" is scientifically meaningless. Postwar social and cultural attitudes toward race changed more slowly, and less decisively, than scientific ones. The popular culture of the late twentieth century reflected this, displaying both the nineteenth-century belief that race defines character and the newer image of race as biologically superficial.

The idea that race defines character persisted in postwar popular culture for a variety of reasons. In some cases, it persisted because of cultural inertia; formula-driven works uncritically recycled stock characters who reflected old ideas about race. In other cases, it persisted by conscious choice. Openly racist whites continued to embrace old assumptions about race defining character, but so did nonwhite activists interested in promoting group pride. Slogans like "It's a black thing, you wouldn't understand" (late 1980s) implied deeply (biologically?) rooted cultural differences. Old ideas about race also persisted subtly but powerfully in political advertising. The infamous "Willie Horton" TV ad used against Democratic presidential candidate Michael Dukakis in 1988 attempted to frighten viewers with an image of prematurely released, implicitly black criminals streaming out of prison to rape or murder again, as the real Horton had while Dukakis was governor of Massachusetts. It used an image as old as American slavery—that of black men as violent, asocial sexual predators hungry for white women. An antipollution public service ad from the early 1970s showed a silent, impassive Indian in beaded buckskin clothes traveling across a litter-strewn landscape and, in the final shot, shedding a single tear. The ad owed its power to the familiarity of the stereotype—an Indian, stoic "by nature," would cry only in a moment of unbearable sorrow.

On the whole, however, the mainstream of postwar American popular culture steadily moved toward treatments of race that separated it from character. The earliest products of this shift, from the 1950s and '60s, are among the best known. "The Sneetches," a 1961 children's story by Dr. Seuss, concerns two groups of birdlike creatures distinguished only by the presence or absence of a green star on their bellies. The star-bellied sneetches, sure of their natural superiority, disdain and exclude their starless brethren, until a mysterious stranger arrives with a machine that can add or remove green stars at will. Seuss's tale elegantly summarizes the emerging idea that our racial differences are trivial when set against our shared humanity. John Ball's adult mystery novel *In The Heat of the Night* (1965) approaches the same territory from a different direction. Its heroes' intense, shared desire to solve a murder forces them to work together, and in the process their relationship moves from mutual contempt to mutual respect. Bill Gillespie (who is white) and Virgil Tibbs (who is black) discover in the end that their souls are more alike than their skins are different.

The 1960s and early 1970s also saw the beginnings of color-blind writing and casting in Hollywood, first in television and later in the movies. Black and, to a much smaller degree, Asian and Indian actors began to appear in parts that could have been played just as easily by white actors. Alexander Scott (Bill Cosby), one of the two heroes of the TV show *I Spy* (1965–1968), was the first such character to achieve real popularity. Others soon followed: Lieutenants Sulu (George Takei) and Uhura (Nichelle Nicholls) in *Star Trek* (1966–1969), electronics expert Barney Collier (Greg Morris) on *Mission: Impossible* (1966–1973), and high school teacher Pete Dixon (Lloyd Haynes) on *Room 222* (1969–1974). *Barney Miller* (1975–1982), a half-hour comedy set in a New York police precinct, marked a watershed in this process. The five wisecracking detectives in its first-season cast, though strikingly multiracial and multiethnic (a black, an Asian, a Latino, a Pole, and a Jew), were written and played with only passing references to their racial and ethnic differences. *Barney Miller* led the way for the color-blind casts that producers like Steven Bochco (*Hill Street Blues*) and David E. Kelley (*Ally McBeal*) made standard in the 1980s and '90s.

Philadelphia, an Oscar-winning 1993 legal drama starring Tom Hanks and Denzel Washington, suggested how much popular culture's portrayal of race has changed. Its plot echoed *In The Heat of the Night*—a black man and a white man, forced together by circumstance, find mutual dislike turning to understanding and respect. Race, however, was beside the point in *Philadelphia*. The two characters' initial distrust was rooted not in the fact that one is white and the other black but in the fact that one is a gay man with AIDS and the other a homophobe.

The same pattern continued through the 1990s and into the early twenty-first century in such movies as *Crimson Tide* (1995) and *Jackie Brown* (1997) and on such TV series as *Homicide* (1994–1999) and *ER* (1994–   ). Race is depicted as one among many character-defining elements—a view with which most scientists would concur.

Related Entries: Evolution, Human; Genes; Intelligence, Human

FURTHER READING

Bogle, Donald. *Prime Time Blues: African Americans on Network Television.* Farrar, Straus, and Giroux, 2001. Rollins, Peter C., and John E. O'Connor, eds.

Hollywood's Indian: The Portrayal of the Native American in Film. University Press of Kentucky, 1999. Two in-depth treatments of how changing attitudes toward race shape, and are shaped by, popular culture.

Gossett, Thomas F. *Race: The History of an Idea in America.* 2nd ed. Oxford University Press, 1997. Originally published in the 1960s; traces the impact of scientific racism in literature and culture.

Gould, Stephen Jay. *The Mismeasure of Man.* Rev. ed. Norton, 1996. Early chapters discuss scientists' attempts to prove the superiority of white to nonwhite races through skull measurements.

Hannaford, Ivor. *Race: The History of an Idea in the West.* Woodrow Wilson Center Press, 1996. Definitive history of scientific racism in Europe and the United States.

Montagu, Ashley. *Man's Most Dangerous Myth: The Fallacy of Race.* 1953. Altamira Press, 1996. Pioneering critique of the idea that race is biologically significant.

Radiation

Radiation is energy, or streams of atomic particles, transmitted through space. It occurs naturally; objects on Earth are continually exposed to radiation from cosmic rays and the decay of radioactive elements in rocks and soil. It is also produced artificially for a variety of human uses: power generation, warfare, microwave cooking, scientific research, medical diagnosis, and the treatment of such diseases as cancer.

Public interest in radiation focuses on its ability to damage or destroy living tissue—a reflection of radiation's close association with horrific images of Hiroshima, Nagasaki, and the ruined nuclear power plant at Chernobyl. Humans exposed to massive doses of radiation often suffer damage to their central nervous systems, corneas, gastrointestinal tracts, reproductive organs, and bone marrow. Those that survive the initial effects of exposure suffer from higher-than-normal rates of cancer, and their children frequently carry genetic mutations caused by the radiation. The effects of lower doses absorbed over longer periods are less clear. This uncertainty has led since 1970 to a series of rancorous public controversies over what constitutes a "safe" level of exposure to radiation.

The stories that popular culture tells about radiation also revolve around its effect on people. Radiation, however, plays different roles in different stories. In some it brings swift and inevitable death; in others it turns the familiar world topsy-turvy.

Radiation naturally appears as an agent of death in stories set in the aftermath of nuclear war. Such stories are invariably meant to be cautionary tales, and in them radiation functions as an invisible Grim Reaper, claiming those who act unwisely. In stories such as *On the Beach* (novel 1957, film 1959) and *Doctor Strangelove* (film 1964), it extinguishes all human life. In slightly more "optimistic" stories, such as *Alas, Babylon* (novel 1959), *Testament* (film 1983), and *The Day After* (TV film 1983), a few survivors are left to piece together the torn fabric of their lives amid the rubble of a ruined world. The message, in both cases, is that peace and arms control are the only sure defenses against radioactive ruin. "Survivalist" fiction, on the other hand, features heroes who survive the war and flourish in the postwar ruins because they have prepared in advance. The genre, which includes both serious novels such as Robert Heinlein's *Farnham's Freehold* (1964) and action-oriented series such as William W. Johnstone's *Out of the Ashes* (1983) and its many sequels, makes radiation a death sentence only for the weak and unprepared. Sometimes it is treated as a kind of blessing, clearing the way for a new and robust society ruled by the competent, the clear-headed, and the strong.

Individual victims of radiation are, in popular culture, heroes who sacrifice themselves to save others rather than (as in after-the-bomb stories) fall victims of their own or their leaders' failings. Astronauts in the movie *Deep Impact* (1998) use a nuclear device to fracture the comet that threatens Earth, knowing they will absorb a fatal dose of radiation in the process. Rhysling, the hero of Robert Heinlein's short story "The Green Hills of Earth" (1947), is a spaceship engineer who risks radiation exposure in order to save his ship on two separate occasions. The first exposure costs him his sight and his job; the second costs him his life. Typical of individual victims of radiation in popular culture, Rhysling dies quickly but not instantly of his injuries. Again typically, he uses the interval between fatal injury and death to make a dramatic final statement. In Rhysling's case, the statement is a song that ensures his place in the folklore of space travel. For most other victims, like the astronauts in *Deep Impact*, it is an emotional farewell to loved ones.

Those that radiation does not kill, it transforms. The results, especially in movies, are often horrific. The giant ants in *Them!* (1954) and the rampaging dinosaurs *The Beast from 20,000 Fathoms* (1954) and *Godzilla* (U.S. cut 1956) are unwanted by-products of nuclear testing. *The Amazing Colossal Man* and *The Incredible Shrinking Man*, both released in 1957, attribute their heroes' size changes to accidental radiation exposures. The subhuman, subterranean mutants featured in *Beneath the Planet of the Apes* (1969) and *Battle for the Planet of the Apes* (1973) have been twisted by radiation in both mind and body. The dissolving flesh and cannibalistic urges of *The Incredible Melting Man* (1977) are legacies of a massive dose of cosmic rays absorbed by the ex-astronaut of the title. Not all radiation-induced changes are for the worse, however. Some of Marvel Comics' best-known superheroes—Spiderman, the Hulk, and the Fantastic Four—fight for truth and justice with superhuman powers created by radiation.

Related Entries: Atomic Energy; Insects, Giant; Mutations; Superhumans

FURTHER READING AND SOURCES CONSULTED

Evans, Kim. *Apocalypse Movies: End of the World Cinema*. St. Martin's, 1999. Includes discussions of how radiation and its effects are shown.

Weart, Spencer. *Nuclear Fear: A History of Images*. Harvard University Press, 1988. Americans' antipathy (not always rational, in the author's view) to both nuclear arms and nuclear power.

Winkler, Allan M. *Life under A Cloud: American Anxiety about the Atom*. 1993. University of Illinois Press, 1999. Excellent survey; less detailed but more neutral than Weart.

Relativity

The theory of relativity, developed by Albert Einstein (1897–1955) in the early twentieth century, consists of two parts: the comparatively straightforward Special Theory and the more complex General Theory. The essence of both parts is that the laws of nature are the same for any observer. The Special Theory restricts that claim to observers at rest or in uniform motion (changing neither speed nor direction); the General Theory extends it to cover all observers. If an observer at rest, an observer in uniform motion, and an observer undergoing acceleration (changing speed or direction) do the same experiments, each will observe the same results and can deduce from them the same laws of nature. There is, Einstein concluded, no privileged vantage point from which to observe the universe. All "frames of reference"—resting, moving, or accelerating—are equally valid for this purpose, and no one of them can be said (as Isaac Newton believed) to represent the "real" or "God's-eye" view.

The basic principles of relativity, described above, are simple. Their effects, however, are complex and staggeringly counterintuitive. Three of the best-known effects, all tied to the velocity of the frame of reference, illustrate this point. First, moving objects increase in mass and contract along the axis of their motion. Second, time passes more slowly in faster-moving reference frames than in slower-moving ones. (Einstein, in one of the many "thought experiments" that he used to explain relativity, illustrated this "time dilation" effect with a pair of imaginary twins. One twin leaves Earth and travels aboard a spaceship at nearly the speed of light. He returns, having experienced the passage of only a few years, but finds his brother an old man and his own children grown to adulthood.) Length contraction, mass increase, and time dilation take place in any moving reference frame, but they become pronounced only at speeds approaching that of light. From this comes a third effect: no object can travel at or above the speed of light, since its mass would become infinite, and time in its frame of reference would slow to a stop. Because the speed of light is constant for all observers in all frames of reference, the effects of relativity make it a universal speed limit.

Relativity's basic principles are a well established, uncontroversial part of modern physics. Their predicted effects have been confirmed many times over by experiment. Their counterintuitive effects, however, make them difficult for nonscientists to grasp quickly or easily. Popular culture, ill equipped to create that understanding and still tell a story, often distills the implications of Einstein's ideas into the aphorism "everything is relative." The aphorism, in turn, is often taken to mean that there are no more absolutes in science or in morality—a position that confuses relativity with relativism. Einstein, along with Sigmund Freud and Karl Marx—creators of psychoanalysis and communism, respectively—thus becomes cast as one of the thinkers who replaced the moral certainties of the nineteenth century with the "anything goes" attitudes that supposedly characterized the twentieth. In fact, he believed in and vigorously argued for the existence of both moral and scientific absolutes. The linkage of relativity and relativism says less about Einstein's ideas than it does about prevailing attitudes at the time, just after World War I, when the public first encountered them (Johnson 1–10).

The countless science-fiction stories that involve travel near, at, or beyond the speed of light should deal with the relativistic effects of such travel, but most simply ignore them or explain them away. Comparatively few, including novels such as Michael P. Kube-McDowell's *The Quiet Pools* (1992) and movies such as *2001: A Space Odyssey* (1968) and *2010: The Year We Make Contact* (1984), acknowledge the limits imposed by relativity and accept the constraints that they place on plot and action. A still-smaller group not only acknowledges the effects of relativity but makes them central to the story. The heroes of Robert Heinlein's young-adult novel *Time for the Stars* (1956) enact Einstein's "separated twins" thought experiment. One of the twin heroes leaves Earth on a starship to serve as one end of a telepathic communication system; he remains young, while his brother, who remains on Earth as the other end of the system, ages sixty years in the course of the story. Poul Anderson's novel *Tau Zero* (1970) provides the ultimate example of time dilation. Its heroes, traveling just below the speed of light, watch the universe age until, in the climax, they witness its death and rebirth.

Related Entries: Einstein, Albert; Space Travel, Interstellar; Speed of Light

FURTHER READING

Calder, Nigel. *Einstein's Universe*. Viking, 1979. One of the best popular treatments of the subject; written by an authority who, unlike Gamow and Einstein, is a nonphysicist.

Einstein, Albert. *Relativity*. 1916. Crown, 1961. Einstein's own introduction, optimistically subtitled "a clear explanation that anyone can understand"; assumes high school math.

Gamow, George. *Mr. Tomkins in Wonderland*. 1939. Reprinted as *Mr. Tomkins in Paperback*. Cambridge University Press, 1993. The fictional title character goes inside Einstein's famous "thought experiments" to illustrate the effects of relativity.

Johnson, Paul. *Modern Times: The World from the Twenties to the Eighties*. Harper and Row, 1983. Chapter 1, "A Relativistic World," describes the vast cultural impact of Einstein's often-misinterpreted theories.

Religion and Science

The relationship between science and religion is complex. It resists easy generalization, varying geographically, chronologically, and from one religious tradition to another. The relationship between science and Christianity in the West—the subject of this essay—is only one facet of a multifaceted story.

Both science and religion are ways of understanding the universe. Both emerged, over the course of the Middle Ages, as significant elements in the worldviews of educated Europeans. Both were studied, taught, and codified at Western universities. Both were promoted, more often than not, by the same people; well into the nineteenth century, many career scientists were also clerics, and many clerics pursued science as a hobby. A thousand years of intimate coexistence has not, however, brought consensus on the proper relationship between science and religion.

Few argue that either science or religion is wholly without value as a source of knowledge. Most stake out their positions somewhere between those extremes, in a middle ground where science and religion can coexist. One popular middle-ground position treats science and religion as two sides of a single coin: God is revealed to humans both through His word (scripture and prophecy) and His works (the natural world). Religion, the study of the word, and science, the study of the works, reveal different aspects of God. Both are essential to understanding Him, and neither can contradict the other. A second argues that seekers of scientific and religious truth can, despite their fundamentally different methods and goals, offer each other useful insights. A third, more austere than either, treats science and religion as wholly separate: science addresses the structure and mechanics of the universe; religion addresses its meaning and purpose. Each is valuable within its area of expertise, but neither should tackle questions that properly belong to the other.

All three views treat science and religion as complementary. The first view holds that science and religion each provide part of the answer to any given question. The second view contends that science deepens religious insights and vice versa. The third suggests that science and religion address complementary sets of questions. All three views see no possibility of genuine conflict between scientific and religious truths, and attribute apparent conflicts to human misunderstanding.

Belief in the peaceful coexistence of science and religion has traditionally appealed to scientists, clerics, and the lay public alike. It eliminates the need to choose between two powerful, attractive ways of understanding the world, each of which is deeply enmeshed in Western society. Popular culture, however, routinely emphasizes conflict in its depictions of science and religion. It treats them as mortal enemies, each inescapably in conflict with the other and capable of advancing only if the other retreats.

According to popular culture, the conflict between science and religion has deep historical roots. The popular image of the Middle Ages as a time of darkness and ignorance rests in part on the belief that the Roman Catholic Church suppressed all attempts at scientific advancement. Rudyard Kipling's 1926 short story "The Eye of Allah" imagines that the microscope was invented in the twelfth century. The prototype winds up in the hands of a Medieval churchman, who contemptuously destroys it, preventing the early arrival of the Renaissance. Fictional time travelers to the Middle Ages—from Mark Twain's "Connecticut Yankee" to Leo Frankowski's Conrad Stargard—often find their attempts to introduce modern ideas blocked by the church or its agents. Novels such as Keith Roberts's *Pavane* (1968) and Kingsley Amis's *The Alteration* (1976) take the idea farther. They are set in alternate worlds, where, because the Protestant Reformation begun by Martin Luther in 1517 failed and Catholic Church's power remained undiluted, the Industrial Revolution did not begin until the mid-twentieth century.

Inherit the Wind, a 1955 play based on the 1925 trial of Dayton, Tennessee, high school teacher John Scopes, takes a similar position. The actual trial was a carefully staged media event, designed by the American Civil Liberties Union to test the constitutionality of Tennessee's newly passed law banning the teaching of evolution in public schools. Scopes volunteered to be charged and tried. The civic leaders of Dayton cooperated eagerly, seeing the trial less as a moral crusade than as a rich source of free publicity. The play paints a much darker picture. The Scopes figure, Bertram Cates, is arrested and jailed against his will. His fellow citizens are an angry mob who sing of hanging him "from a sour apple tree," and the chief prosecutor, Matthew Harrison Brady, is a blustering demagogue. Brady and the mob are written (and, in the famous 1960 film adaptation, played) as self-righteous fanatics, drunk on their own piety. Henry Drummond, the play's wise and rational lawyer-hero, must save Cates—and, by extension, all Americans—from Brady's misguided attempts to halt scientific progress.

Portrayals of past encounters between science and religion invariably cast the scientist as the hero and the church (or its representatives, like Brady) as the villain. Portrayals of present-day encounters are more complex. Religion is not necessarily evil, and science not necessarily good. Nor are they necessarily represented by different characters. The assumption that science and religion must conflict, not cooperate, remains intact.

Ellie Arroway, the scientist-hero of Carl Sagan's *Contact* (novel 1985, film 1997), is skeptical or dismissive of religion and devout believers for most of the story. Indeed, she has good reason to be. Religious fanatics first protest and then sabotage her attempts to communicate with an advanced alien civilization, and an ultraconservative religious advisor to the president thwarts her efforts to gain government support. A question about her own religious beliefs, which she answers honestly, derails her effort to be Earth's first emissary to the aliens. Over the course of the story, however, her developing relationship with unorthodox religious leader Palmer Joss forces her to reexamine her resolute nonbelief. The story's climax leaves her in a position unfamiliar to scientists but familiar to the faithful—asking others to accept without evidence what she knows in her heart to be true.

The tensions between science and religion are sharper in Arthur C. Clarke's short story "The Star" (1955), and the emotional stakes are higher. The first-person narrator is both an astronomer and a Jesuit priest, working aboard a spaceship exploring distant star systems. The ship stops to explore the remains of a long-dead alien civilization, destroyed when its sun went nova (exploded). The astronomer, deeply immersed in his science and reveling in the precise certainty of the answers it provides, calculates the date and position of the nova as it would have been seen from Earth. Only after the calculations are complete does he realize their implications. He ends the story with his dual faith in science and Christianity intact, but in the final sentence asks God: "What was the need to give these people to the fire, that the symbol of their passing might shine above Bethlehem?"

Related Entries: Darwin, Charles; Galileo; Intelligence, Human; Life, Origin of

FURTHER READING AND SOURCES CONSULTED

Barbour, Ian G. *Religion and Science: Historical and Contemporary Issues*. Rev. ed. HarperCollins, 1997. A comprehensive survey by a leading science-and-religion scholar; perhaps the best one-volume treatment available.

Brooke, John Hedley. *Science and Religion: Some Historical Relations*. Cambridge

University Press, 1988. A compact historical overview; aimed at undergraduates but argued and written at a fairly sophisticated level.

Larson, Edward J. *Summer for the Gods: The Scopes Trial and America's Continuing Debate over Science and Religion.* Harvard University Press, 1997. Pulitzer Prize–winning history of the Scopes trial and its place in American culture, including its relation to *Inherit the Wind.*

Polkinghorne, John. *The Faith of a Physicist: Reflections of a Bottom-Up Thinker.* Fortress Press, 1996. A respected physicist who is also an Anglican priest argues for the natural intertwining of science and religion.

Ruse, Michael. *Can a Darwinian Be a Christian?* Cambridge University Press, 2000. A distinguished philosopher of science (and committed Darwinist) takes on the title question and answers "Yes!"

Reproduction

Human reproduction, little changed for thousands of years, has been transformed by science and technology since 1960. The once-inseparable acts of sex and reproduction can now, for people with sufficient resources, be completely separated from one another.

The most familiar face of this revolution is chemical birth control. Introduced in the early 1960s, its ease of use and high (99 percent) effectiveness rate gave women who had access to it unprecedented control over when and if they would bear children. Birth-control pills completed a process begun by condoms and diaphragms—the separation of sex from reproduction. It accelerated existing social and cultural trends: smaller families, greater autonomy for women, and the relaxation of cultural limits on sexuality.

The impact of birth control pills was hailed in Loretta Lynn's hit country song "The Pill" (1974) and in women's lifestyle magazines (notably *Cosmopolitan*). Over time, however, it has grown steadily less visible in popular culture. Fictional characters rarely discuss birth control, even though censors would permit it and sex educators would welcome it. The reason reflects the Pill's impact on American society—chemical birth control is now readily available to any independent adult woman with a middle-class income (that is, nearly every sexually active woman in American popular culture). Audience members thus assume without being told that the woman who just tumbled joyously into bed with her lover is using them. Birth control has, for such women, become routine. Like all routines—flossing teeth, folding socks, feeding the cat—it disappears from the lives of fictional characters unless its presence is important to the story.

The other face of the revolution involves the opposite process, enhancing fertility and promoting conception. Fertility drugs have, in recent decades, given otherwise infertile couples opportunities to bear children. Infertility cases for which drugs alone would be ineffective can increasingly be treated by other means. In-vitro fertilization, which had its first success in 1978 with the birth of Louise Brown in England, is now a standard (though costly) procedure. Conception using donated sperm or eggs also became more common in the 1980s and 1990s. The use of surrogate mothers to carry an embryo to term, and the freezing of eggs and sperm for later use, are still uncommon, but unlikely to remain so, in the early twenty-first century.

Births facilitated by in-vitro fertilizations, donated eggs, or surrogate mothers are still comparatively rare in popular culture, because they are still far from routine in the real world. The techniques involved are well tested but useful to only a small segment of the population. They are *available* only to an even smaller group, those with ample financial resources and access to advanced medical care. When medically assisted conceptions do appear in popular culture—whether in documentaries, newspaper and magazine feature stories, or made-for-TV movies—they are generally treated in narrow, highly personalized terms. The stories, whether fact or fiction, nearly always follow the same dramatic arc— "John and Jane go to great lengths to try to have a baby of their own." This personalization virtually demands that the audience take John and Jane's side. They're nice people, dealt a bad hand by nature. Who would oppose their attempts to improve that hand and have the baby they want so much? Skepticism about the process in general is hard to maintain in the face of a well-told story about a *particular*, appealing couple.

Popular culture's treatment of assisted conception reflects a basic cultural convention, that children are a blessing and enforced childlessness a tragedy. Limits on reproduction imposed by anyone *but* the prospective parents are, in popular culture, evil and unjust by definition, and stories about them often pit an appealing, childless couple against a heartless government. Films like *Zero Population Growth* (1972) and *Fortress* (1993) feature young parents who become fugitives after conceiving a child forbidden to them by law. The more thoughtful *Gattaca* (1997) makes its hero a victim of genetic apartheid—a "love child" conceived outside the government's efforts to produce genetically perfect citizens. Aldous Huxley's novel *Brave New World* (1932) offers an even more radical vision—a world where parenthood itself has been abolished and babies are grown and raised in laboratories to state specifications.

P.D. James's novel *The Children of Men* (1993) is the ultimate commentary on enforced childlessness. It posits that all human males have become suddenly and permanently sterile and chronicles the first years of humankind's downward spiral toward extinction. Nature, in James's world, has dealt humankind the ultimate losing hand, one that science is powerless to improve.

Related Entries: Cloning; Evolution, Human

FURTHER READING

Diamond, Jared. *Why Is Sex Fun? The Evolution of Human Sexuality.* Basic Books, 1998. A brief, topical introduction to the evolutionary roots of human sexuality and reproductive patterns.

Silber, Sherman J. *How to Get Pregnant: With the New Technology.* Warner Books, 1998. A detailed, comprehensive resource book that sees technology-assisted reproduction as a boon.

Tane, Andrea. *Devices and Desires: A History of Contraception in America.* Hill and Wang, 2001. Emphasizes the diversity of contraceptive options before the Pill, and the role of public demand in promoting new ones.

Robots

A robot is a mechanical device designed to do work ordinarily done by humans. Robots typically resemble the human beings they replace only when that allows them to function more efficiently. Robotic arms used on automobile assembly lines have "shoulder," elbow," and "wrist" joints that allow them to mimic a human arm's range of motion. Robots designed to move in warehouses and other hard-floored environments, on the other hand, use wheels rather than human-style legs and feet. All robots are, at some level, controlled by human operators; the control may be direct (through inputs from a remote console) or indirect (through instructions programmed into a computerized "brain"), but it is always present. Even robots endowed with artificial intelligence and capable of adjusting to their environment as they "learn" about it depend on human input to define the basic patterns of their responses.

ROBOTS IN THE REAL WORLD

Advances in computers and miniaturized electronic components made robots increasingly capable, and so increasingly attractive to potential users, over the course of the late twentieth century. Robots are now widely used for industrial work that requires the precise repetition of a pattern, rather than flexible human judgement. Robot arms, for example, can weld automobile components more quickly, more precisely, and more consistently than all but the best human welders. Robots are also widely used in place of humans for the exploration of dangerous environments: the inside of damaged nuclear power plants, the depths of the sea, and the surfaces of other planets. *Sojourner,* a small robotic rover landed on Mars by the *Mars Pathfinder* mission, became an overnight celebrity in the summer of 1997. Robots are also, increasingly, the first observers sent into potentially violent situations. They examine sus-

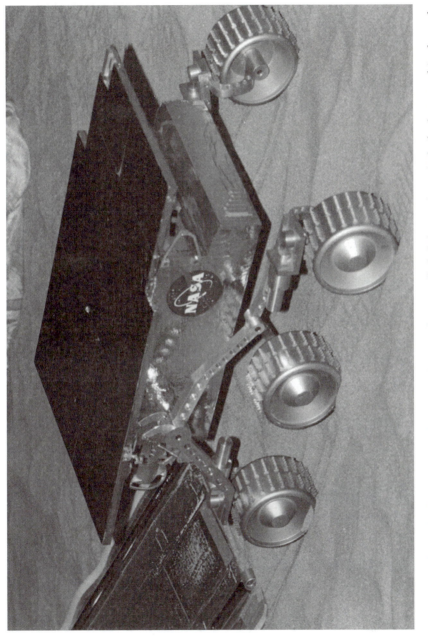

Sojourner on Mars, 1997. Compact, wheeled, and remotely controlled, *Sojourner* is typical of robots used in the real world to explore hazardous environments. Courtesy of NASA/JPL/CalTech.

pected bombs, investigate buildings where hostages are held, and survey the skies over battlefields—all without putting humans at risk.

Virtually all the robots now in operational use have been designed with specific tasks in mind. They operate in known environments, under known conditions, confronting problems that their designers and operators have thought through in advance. However, general-purpose robots, capable doing a wide range of tasks and operating in difficult or unknown environments, are also being developed. They pose enormous challenges to designers, and their success will, almost inevitably, depend on the development of artificial intelligence at a level far more sophisticated than any yet reached.

FICTIONAL ROBOTS: NONHUMANOID

Fictional versions of the robots used or projected for use in the real world are comparatively rare in popular culture. Occasionally, as in the real world, they are simply tools, whirring away in the technological background of the story. Military adventure novels such as Dale Brown's *Shadows of Steel* (1996) and James Cobb's *Sea Fighter* (2000), for example, present automated reconnaissance aircraft (known as "unmanned aerial vehicles") as one intelligence-gathering tool among many. More often, fictional robots act like mechanized versions of animals. "Bad" robots (like the bears and wolves of folklore) blindly oppose the hero and represent another obstacle to overcome. "Good" robots (like the dogs and trained horses of folklore) offer the hero aid and comfort, often exhibiting considerable intelligence in the process.

Bad nonhumanoid robots are seldom actively, consciously evil. They can no more override their programming than an animal can override its instincts. The automated defense systems featured in the climax of *The Andromeda Strain* (novel 1970, film 1971) do not realize that the escaped "animal" they are trying to kill is really a human scientist. The insectlike robots in movies like *Runaway* (1984) and *Star Wars: The Phantom Menace* (1999) are blindly following the instructions of their evil creators. "Amee," the canine robot in *Red Planet* (2000), hunts down the human members of its crew because the damage she sustained in a crash landing has locked her into "combat mode."

Good nonhumanoid robots are, to the extent that a metal-and-plastic object can be, cute and cuddly. They have personalities instead of just instincts, and they are deeply devoted to their human masters. Huey, Louie, and Dewey, the three robot sidekicks in the 1971 film *Silent Running*, seem far more "human" than most of the human characters. In the *Star Wars* saga, Luke Skywalker relies on R2D2 in ways the way that Lassie's many fictional owners relied on the famous collie. All that changes is the dialogue; "Lassie, get help!" becomes "Shut down all the garbage mashers on the detention level!"

FICTIONAL ROBOTS: HUMANOID

Fictional robots are constrained neither by the current limitations of mechanical components nor by the current limitations of artificial intelligence. They *are* constrained, however, by their creators' need to tell engaging stories. They tend, as a result, to be more capable and more "human" than their real-world counterparts—predominantly vertical lines, with bilateral symmetry, paired appendages, and sensory equipment grouped at the top of the "body." Even if they move on wheels or treads rather than feet, they frequently have two separate "legs," with joints in the same places as human ones. The *Star Wars* saga's tripodal, barrel-shaped R2D2 is a more plausible robot design than his humanoid partner C3P0, but it is "Threepio" (looking like a high-tech version of the Tin Man from *The Wizard of Oz*) that better represents the typical robot of popular culture.

Fictional robots also act in distinctly human ways. They can speak idiomatic English, solve complex problems, and think creatively about data they have never before encountered. Most important, perhaps, they have humanlike personalities. The never-named robot of the TV series *Lost in Space* (1965–1968) is both physically and morally protective of young Will Robinson. The robots in Isaac Asimov's stories and novels, governed by the commandment-like "Three Laws of Robotics," face moral dilemmas and suffer intense psychological stress in trying to resolve them. The robot assistant who alternates with his human master as narrator of Alan Dean Foster's detective story *Greenthieves* (1994) is given to tart-tongued observations about the human characters. The humanization of robots is most telling, however, when it involves less-than-admirable traits. *Star Wars'* C3PO is fussy and self-important; the "robot pilot" deployed to save a damaged airplane in the Bugs Bunny cartoon "Hare Lift" (1952) quickly sizes up the situation and bales out with the last remaining parachute. Humanoid robots are, all in all, neither better nor worse than their flesh-and-blood creators, only different.

It is that difference, of course, that makes robots interesting as dramatic devices. Simultaneously human and not human, they can comment on humans from an outsider's perspective or explore what it means to be human. The robot in *Lost in Space*, for example, served as the voice of reason and moral authority in stories focusing on Will Robinson and the nefarious stowaway Dr. Zachary Smith. "Steel," a 1963 episode of *The Twilight Zone*, is a study of two characters: a robot prizefighter named "Battling Maxo" and his human manager, an ex-fighter compelled by his sense of honor to take Maxo's place in a bout against another robot. Eric

Frank Russell's short science-fiction story "Jay Score" (1941) explores the relationship between a spaceship pilot and the shipmates he saves, at great cost to himself, during an in-flight emergency. Only the last line of the story reveals that the heroic pilot is actually a robot.

FICTIONAL ROBOTS: GOD-MACHINES

Most fictional robots have capabilities qualitatively similar to those of humans. The things they can do are things that humans can also do, though the robots may well do them better, faster, or with more uniform results. Popular culture occasionally, however, also features robots whose capabilities exceed those of humans not in degree but by entire orders of magnitude. These robots, while clearly mechanical, are powerful enough to seem godlike. Their roles in the story are, in fact, often similar to those traditionally played by gods. When they threaten humans, they are apparently unstoppable; when they intervene in human affairs, they can effortlessly change the course of history. Comparing them to "forces of nature," though superficially inaccurate, captures their vast power.

Gort, the nine-foot-tall robot who guards alien ambassador Klaatu in the film *The Day the Earth Stood Still* (1951) is a modest example of the type. Bulletproof and amazingly strong, he is "only" capable of vaporizing tanks and rifles over line-of-sight distances. The title character of Ted Hughes's 1968 novella "The Iron Man" (filmed in 1999 as *The Iron Giant*) is also modestly powerful, as are the Autobots and Decepticons of the *Transformers* TV series. "V'ger," the enigmatic alien robot whose origins and intentions drive the plot of *Star Trek: The Motion Picture* (1979), is a more impressive specimen. The opening scenes of the movie show it casually destroying a large space station and a pair of Klingon warships. It is, like most godlike robots, the product of an advanced alien civilization.

The ultimate godlike robots, however, are capable of demolishing not just tanks or spacecraft but entire planets. Fred Saberhagen's "Berserker" stories and novels (1967–) follows humankind's (barely) successful resistance to fleets of such weapons, programmed to exterminate biological life. "The Doomsday Machine," a 1967 episode of the original *Star Trek* TV series, focuses on a battle with a single robot that carves up planets with an energy beam. The machine is (again) beaten, but only by a most drastic measure—using a thousand-foot-long starship as a flying bomb. Greg Bear's 1987 novel *The Forge of God* comes to what is, unfortunately, a more plausible climax: robotic weapons deployed by an unknown alien species literally tear the earth apart. Interestingly, these unstoppable machines display precisely the qualities that make industrial robots valuable in the real world: great efficiency, unshakable focus, and slavish execution of their makers' instructions.

Related Entries: Androids; Cyborgs; Intelligence, Artificial

FURTHER READING AND SOURCES CONSULTED

D'Aluisio, Faith, and Peter Menzel. *Robo Sapiens: The Evolution of a New Species.* MIT Press, 2000. Discusses current efforts to build intelligent humanoid robots.

Maravec, Hans. *Robots: Mere Machine to Transcendent Mind.* Oxford University Press, 2000. Extravagant vision, rooted in current work, of a future where, by 2050, robots will exceed human intelligence and replace humans.

Robot Information Central. Arrick Robotics. 18 May 2001. <http:// www. robotics.com/robots.html>. An enormous, categorized collection of pages and outside links covering the history, current state, and applications of robotics. Source of the real-world technological information in this entry.

RobotCafé.com. 17 May 2001. RobotCafé.com. 18 May 2001. <http:// www. robotcafe.com/>. Over 500 links to robotics resources in print and online, including up-to-date robotics news; aimed, slightly more than Robot Information Central, specifically at robot builders.

Telotte, J.P. *Replications: A Robotic History of the Science Fiction Film.* University of Illinois Press, 1995. Analysis of major robot and android characters in science-fiction films, focusing on the use of them to make points about humans.

Scientific Theories

Humans do not have complete knowledge of the natural world or of any part of it. Knowledge depends on observation, but our powers of observation are limited, and machines can only enhance them so far. The best we can do is to study the pieces we *can* observe and try to infer from them the larger patterns of which they are a part and the underlying forces that create those patterns. Scientific theories are our attempts, inevitably imperfect, to sketch in those patterns and processes based on what we know at a given moment in time. Acquiring new knowledge, or viewing old knowledge from new perspectives, can change what we "know" and lead us to sketch the patterns and processes differently. Scientific theories are, therefore, works continually in progress.

Theories exist to explain sets of observations. They stand or fall, in the long run, on the basis of how well they do so. Successful theories—Newton's universal gravitation, for example—win support by accounting for what is already known of the phenomena they explain *and* predicting discoveries yet to be made. Once-successful theories—Aristotle's Earth-centered universe, for example—lose support when observations that they cannot explain become too numerous or too striking. Neither process takes place overnight, however, and neither takes place in a vacuum. Old theories with waning explanatory power may persist if they complement existing theories, flatter popular prejudices, or have no rivals. New theories with superior explanatory power may languish if their predictions are limited, their supporting evidence shaky, or their social implications unsettling.

"Theory," in everyday usage, is a synonym for "opinion" or "educated guess." It often highlights a sharp contrast between belief and reality. Sentences that begin "Theoretically . . ." or "In theory . . ." typically end with a statement that the real world did not (or is not expected to) conform to the speaker's expectations. Popular culture routinely treats scientific theories as if they were theories in the everyday sense of the word—"guesses" that have little connection to the real world and are likely to crumble when brought into contact with it. Knowledge, in popular culture, falls into one of two categories, "fact" or "opinion" (of which scientific theories are a subset). All facts are equally true, and all opinions equally suspect.

Popular culture's artificial dichotomy between fact and theory is most visible in confrontations between heroes armed with facts and villains who cling blindly to theories. The heroes' view of the world summarizes what they have directly observed. They explain it, only to be rebuffed by the villains on the grounds that it clashes with what the villains "know" to be true. Later developments show that the villains' theoretical "knowledge" is a tissue of assumptions, inferences, and suppositions. The heroes triumph by actively demonstrating facts that expose, in spectacular fashion, the limitations of the villains' theories.

Fictional versions of the story are common. One variation involves a hero who finds a lethal force or creature but cannot persuade theory-bound authorities to accept its existence. Carl Kolchak, the reporter-hero of TV's *Kolchak: The Night Stalker* (1974–1975), never persuades his editor to accept the *real* (supernatural) story behind the grisly deaths he investigates. Fox Mulder of TV's *The X-Files* (1996–) takes years to persuade his partner, Dana Scully, that there might be forces in the universe that her scientific worldview does not recognize. Oceanographer Matt Hooper uses his firsthand knowledge of sharks to puncture elected officials' naïve beliefs in *Jaws* (novel 1974, film 1975), but his own theories about sharks (and, in the book, Hooper himself) are torn apart in the climactic battle. A second common variation of the story involves heroes who save the day by trusting hard-won experience rather than theoretical prescriptions. Maintenance chief Joe Patroni clears a vitally needed runway in *Airport* (novel 1968, film 1970) by gunning the engines of a snowbound jetliner until it breaks free. The operations manual, written by theory-bound engineers, insists that such abuse will destroy the plane; Patroni, in command of the facts, knows better.

Versions of the story featuring historical figures are less common but endlessly retold. Galileo, according to legend, exposed the fallacy of Aristotle's system of physics by dropping iron balls from the leaning tower of Pisa. Columbus, again according to legend, used maps, globes, and the success of his transatlantic voyage to demolish the supposedly widespread belief that the earth was flat. Accounts of the Wright Brothers' early flights are often juxtaposed with statements by contemporary physicists—Lord Kelvin in 1892, Simon Newcomb in 1902—that heavier-than-air flight is impossible. The space age has its own version of the story—that in 1922, professors at the University of Munich rejected Hermann Oberth's dissertation on interplanetary rockets because in space "they would have nothing to push against." Their view too is routinely juxtaposed with facts (images of rockets taking off) that prove it spectacularly wrong.

Finally, stories of conflicts between fact and theory are central to the rhetoric of "anomalists"—advocates of ideas, such as ancient astronauts, creationism, ESP, and "pyramid power," that are dismissed by mainstream science. Anomalists make two basic assumptions that mainstream scientists reject: that "real" scientific knowledge emerges directly from the facts, and that even a handful of contrary facts can overturn even a well-established theory. They cast themselves as the keepers of such facts, and so as the heroes in the traditional story. Mainstream scientists thus become the villains of the story, too wedded to their theories to see the truth that the facts represent. Facts that anomalists find especially potent—artifacts, photographs, narratives, or even sets of numbers—often take on talismanic qualities in their eyes. "This," their rhetoric often implies, "is the magic fact that will slay the dragon of mainstream theory."

The dramatic appeal of such confrontations is easy to understand. They tap into a deep Western fondness for stories about the clever individual who makes fools of the learned authorities; Galileo, dropping his iron balls off the tower in Pisa, is like the boy who proclaims that the emperor has no clothes. The confrontations also tap into a specifically American fondness for tales that celebrate egalitarianism—the working-class maintenance man of *Airport* knows more about "his" airplanes than the college-educated engineers who designed them. Americans, as a group, have always tended to denigrate abstract knowledge and those who possess it. "Bookworm," "egghead," and "ivory tower intellectual" are terms of derision; the inventor, not the physicist, is the quintessential American hero. American popular culture's depiction of theories and theorizing reflects this prejudice. Theory, in popular culture, is the province of self-absorbed, self-important fools. Facts are accessible to anyone, including the practical hero who uses his command of them to save the day and show the theorist how the *real* world works.

Related Entries: Experiments; Ideas, Resistance to; Intelligence, Human

FURTHER READING AND SOURCES CONSULTED

Bauer, Henry H. *Scientific Literacy and the Myth of the Scientific Method*. University of Illinois Press, 1994. Discusses the interaction of scientific theory and social context.

———. *Science or Pseudoscience: Magnetic Healing, Psychic Phenomena, and Other Heterodoxies*. University of Illinois Press, 2001. Contrasts approaches of anomalists and mainstream scientists.

Giere, Ronald N. *Understanding Scientific Reasoning*, 4th ed. Holt, Rinehart, and

Winston, 1998. Includes an accessible, rigorous philosophical discussion of theory.

Kuhn, Thomas S. *The Structure of Scientific Revolutions*. 2nd ed. University of Chicago Press, 1970. Landmark study of how scientific theories change over time; principal source for the view of mainstream science in this entry.

Sharks

Sharks are among the oldest and most anatomically primitive of all the fishes. They have jaws and gills, but their skeletons are made of cartilage rather than bone. The basic anatomical features of sharks—their sleek shape, pointed or shovel-like snout, and distinctive dorsal fin—have changed very little in nearly 200 million years. Sharks are among the most efficient predators in the sea. They have few natural enemies except humans, who have reduced the populations of many shark species to dangerously low levels.

More than 350 species of sharks now inhabit the world's oceans and rivers, and they vary widely in size, appearance, and habits. The smallest sharks are little more than a foot in length, the largest over fifty feet. The very largest species, whale sharks and basking sharks, eat plankton that they strain from the water, as baleen whales do. Many small and medium-sized species are bottom feeders, their teeth adapted to crushing rather than tearing. Most familiar shark species—blue, mako, tiger, hammerhead, and Great White—feed on fish and marine mammals. Their sensory organs allow them to detect blood and thrashing motions in the water—signs of wounded, easily caught prey. These signals, combined with poor eyesight or (rarely) innate aggression, lead to 75–100 shark attacks on humans each year, ten to twenty of which prove fatal.

Popular culture's portrayal of sharks is resolutely one-dimensional and resolutely negative. Even among predators, this is a rare distinction. Lions, tigers, and bears all get depicted positively from time to time. The much-hated wolf is noble in Rudyard Kipling's *Jungle Book* (1894), and spiders are heroes in E.B. White's *Charlotte's Web* (1942) and Roald Dahl's *James and the Giant Peach* (1961). With the lone exception of *Jabberjaw*, an undistinguished cartoon series that ran from 1976 to 1977, sharks are *never* the "good guys." The word "shark" is slang for ruthlessly predatory humans—"pool shark" or "loan shark," or Harvey Mackey's bestselling business book *Swim with the Sharks* (1988). A joke has a lawyer asking why, in the aftermath of a shipwreck, he was the only person not eaten by sharks. The sharks' reply: "Professional courtesy!"

Sharks that aggressively prey on humans, rare in nature, are the norm in popular culture. It presents them by implication as representative of all sharks and so encourages audiences to see any shark that appears in a scene as a clear and present danger to the human characters. Sea stories from Herman Melville's *Moby-Dick* (1851) and Jules Verne's *Twenty Thousand Leagues under the Sea* (1869) to Peter Benchley's *The Deep* (1976) use them as sources of menace. So too do sea-oriented films like *Shark!* (1969), the James Bond adventure *The Spy Who Loved Me* (1977), and the Disney Studios version of *The Little Mermaid* (1989). Sharks also figure prominently in other illustrations of peril at sea. Winslow Homer's painting *The Gulf Stream* (1899) shows them circling a lone sailor clinging to the wreckage of his boat.

The sharks depicted in these instances are ordinary, generic sharks. They are dangerous to humans, but only to humans who venture unprotected into the water. Documentary films from 1971's *Blue Water, White Death* to those featured on the Discovery Channel's annual "Shark Week" use that convention to generate drama. Their diver-cinematographers, protected by steel cages, can shoot from the prey's point of view without becoming prey themselves. *Jaws* (novel 1974, film 1975) terrified its audiences by carefully breaking the convention. The twenty-five-foot Great White shark in *Jaws* begins with a traditional victim—a young woman swimming alone, at night, from a deserted beach. Soon, however, it becomes more ambitious; it kills a boy on an inflatable raft, sinks a small fishing boat, and finally destroys the specialized shark-hunting boat sent to hunt it down. One of the heroes goes underwater in a shark cage, only to have it torn apart by the monster.

The sharks in the inevitable movie sequels to *Jaws* are even more aggressive. *Jaws 2* (1978), for example, shows a Great White leaping from the water in order to pull a hovering helicopter to its doom. Post-*Jaws* horror stories about sharks have upped the dramatic ante even further. The genetically enhanced sharks in the movie *Deep Blue Sea* (1999) pursue humans through a disintegrating undersea habitat. Thomas Alten's novels *MEG* (1997) and *The Trench* feature a giant, prehistoric shark called *megalosaurus* that is capable of attacking submersibles and other high-tech cocoons for fragile humans.

Stories such as *Deep Blue Sea* and *MEG* are clearly influenced by *Jurassic Park* (novel 1990, film 1992) and *Terminator 2* (1991)—thrillers about seemingly invincible predators. Sharks are, however, uniquely suited to such stories because of the image popular culture has created for them—that of a relentless, instinct-driven "eating machine."

Related Entries: Dolphins; Intelligence, Animal; Whales

Detail from *The Gulf Stream* (1899), by Winslow Homer. The sharks' aggressive poses owe more to artistic license than biological reality. Courtesy of the Library of Congress.

FURTHER READING

Allen, Thomas B. *The Shark Almanac.* Lyons Press, 1999. Summarizes the biology and behavior of 100 representative species.

Mathiessen, Peter. *Blue Meridian: The Search for the Great White Shark.* Penguin, 1998. An account, originally published in the early 1970s, of a pioneering expedition to film Great Whites under water.

Taylor, L.R., et al. *Sharks and Rays: A Nature Company Guide.* Time-Life Books, 1997. A comprehensive reference for lay readers, with excellent illustrations.

Space Travel, Interplanetary

Interplanetary travel—that is, travel within our own solar system—poses major technological and operational challenges. It is far less daunting, however, than interstellar travel. Interplanetary travel is already within our grasp, and significant progress within the next century is virtually certain. Interstellar travel—except in its most basic, message-in-a-bottle form—is well beyond our grasp at the moment and likely to remain so. The crucial difference between the two is the distances involved. The moon, the closest planet-sized body to Earth, is only a quarter-million miles away, a journey of a few days with the chemical rockets developed in the 1960s. Neptune, one of the most distant, is 2.8 billion miles away on average, a journey of nearly a decade with chemical rockets and a carefully planned trajectory. Alpha Centauri, the nearest star, is 27 trillion miles distant. A spaceship traveling at a million miles per hour—forty times faster than those now in use—would take 3,000 years to make the trip (Brennan 6). Our own solar system will, for the foreseeable future, be the only space that humans can explore, either in person or by robot proxy.

INTERPLANETARY TRAVEL IN THE REAL WORLD

Our solar system consists of nine planets, scores of moons, hundreds of large asteroids, thousands of smaller ones, and an unknown number of comets. More than a 100 robot spacecraft have, since the beginning of the Space Age in 1957, investigated eight of the nine planets (far-distant Pluto is the exception) and a number of the larger moons. Human astronauts have, by contrast, left Earth's gravitational pull only nine times: three flights around the Moon and six landings on it, all made between December 1968 and December 1972 as part of Project Apollo. No new interplanetary flights with human crews are officially contemplated,

Apollo 15 astronaut and lunar rover, 1971. Project Apollo's six moon landings are, as of this writing, humans' only firsthand explorations of another world. Courtesy of the National Space Science Data Center and Dr. Frederick J. Doyle, Principal Investigator, Apollo 15.

though planning commissions have endorsed them and space enthusi-asts have written book-length proposals for them.

The gap between robot and human exploration of other planets reflects the relative difficulty of the two types of mission. The Apollo missions to the moon spent only a few days in transit, but spacecraft using chemical-fuel rockets would take years to reach other planets. The equip-ment, supplies, and living space necessary to sustain a human crew for years add expense and complexity to the spacecraft. They demand a larger, heavier spacecraft, which in turn requires more-powerful engines and more fuel to drive it. Robot spacecraft can be smaller, simpler, and cheaper. They have, therefore, perennially appealed to cash-starved space agencies like NASA.

INTERPLANETARY TRAVEL IN POPULAR CULTURE: THREE MODELS

Interplanetary flight has, for the four decades since it became possible, been undertaken only by wealthy, technologically sophisticated nations. The staggering expense of the spacecraft, facilities, and support systems involved limits participation to groups with enormous financial re-sources. The commercialization of space travel in and to Earth orbit, in contrast, is already being planned and may become firmly established later in the twenty-first century. The commercialization of interplanetary travel is another matter; the absence of any financial return in the fore-seeable future discourages corporate involvement. Unless economic op-portunities of great value—enough to offset the enormous costs of interplanetary travel—are discovered on other worlds, commercial in-terplanetary travel is unlikely to develop.

Popular culture's depictions of interplanetary travel are more optimis-tic. They take interplanetary travel by human crews for granted and accord robots little, if any, independent role. Popular culture's depictions are also more complex, involving three different, sometimes overlapping, models of the future of interplanetary travel. The first extrapolates the realities of present-day interplanetary travel, such as the Apollo pro-gram, into the future. The second assumes that interplanetary travel will develop commercially, as rail travel did in the nineteenth century and air travel did in the twentieth. The third posits a future in which small groups or even individuals can build their own spacecraft and visit other worlds.

The Apollo Model: Expensive, Slow, and Rare

Stories of interplanetary travel that strive for realism generally assume that the future of interplanetary flight will be much like its past. The ships in these stories may be larger and more sophisticated than Apollo-era spacecraft, but they are only marginally faster. They have nuclear engines, powerful computers, and (often) artificial gravity—but not the kind of next-generation propulsion systems that would reduce inter-planetary travel times from months to days. They are built by govern-ments and, even while in space, operate under the close supervision from the ground—the model used by both the Soviet and American space programs since the 1960s.

Stories about the exploration of Mars, especially, often use this model. Its basic outlines were visible in such novels as Arthur C. Clarke's *The Sands of Mars* (1954) and films like *Robinson Crusoe on Mars* (1964), even before actual attempts at interplanetary flight. The aborted Mars mission in the film *Capricorn One* (1978) is to be carried out using Apollo hard-ware. One publicity still shows an Apollo "lunar" module standing on the familiar rocky plains of the Red Planet. More recent stories about Mars expeditions follow even more closely the Apollo-era model of slow flights, big ships, and a strong government presence. Ben Bova's novels *Mars* (1992) and *Return to Mars* (1999), as well as the film epics *Mission to Mars* (2000) and *Red Planet* (2000), all use it.

The most striking examples of the Apollo model in fiction, however, are the voyages from Earth to Jupiter in the films *2001: A Space Odyssey* (1968) and its sequel *2010: The Year We Make Contact* (1984). Both imprint the basic outlines of Apollo-era space travel on their imagined futures. Their flights are big, expensive projects using cutting-edge technology and astronauts who are firmly under the control of political leaders on Earth. Both also have the "look and feel" of real-world space travel of the decades in which they were made; *2001* replicates the sterile cabins and tight-lipped astronauts of the 1960s, while *2010* recalls the lived-in spacecraft and pilot-scientist teams that characterized space shuttle flights in the early 1980s.

The Railroad Model: Expensive, Fast, and Common

Imagined futures in which interplanetary flight is fast and routine al-low a wider range of stories. Spaceships become in such stories the equivalent of trains in a Western or airliners in an espionage thriller—occasionally the focus of the action but usually just a convenient way of moving characters around. The spaceships in such stories are usually (but not always—see below) large, reliable, and well appointed. They are—like trains or airliners—owned and operated by large companies that turn a profit, despite enormous overhead costs, by providing fast and reliable access to otherwise inaccessible places.

This view of interplanetary travel figures prominently in Robert A. Heinlein's "Future History" stories, written between 1939 and the early

1950s and collected in *The Past through Tomorrow* (1967). The spaceships that link Earth to its off-world colonies have regular routes, regular schedules, and smoothly efficient crews whose principal loyalty is to the company. D.D. Harriman, the founder of commercial space flight, is patterned after real-world innovators in commercial transportation; he shares a last name with E.H. Harriman of the Union Pacific Railroad and a swashbuckling-capitalist persona with Juan Trippe of Pan-American Airways. Heinlein's universe also has its own Casey Jones—an engineer named Rhysling who, in "The Green Hills of Earth" (1947), saves his ship and passengers at the cost of his own life by staying at his post in an engine-room emergency.

The Lone Inventor Model: Cheap, Fast, and Common

Not all interplanetary spaceships in popular culture are big, elaborate, or expensive. Inventors who travel to other worlds in home-built spacecraft appeared decades earlier than their government- and corporate-sponsored brethren, in stories such as Jules Verne's *From the Earth to the Moon* (1865) and H.G. Wells's *First Men in the Moon* (1901). The basic premise of such stories is that space travel is easy, that any talented inventor or skilled mechanic can build and fly a spacecraft capable of reaching other worlds. The stories are especially popular in the United States, where their celebration of individual initiative and the value of practical know-how reinforce deeply held values.

American stories about home-built spacecraft are, because of this resonance, less about nuts-and-bolts accuracy than about the American character. Robert Heinlein's young-adult novel *Rocket Ship Galileo* (1947), for example, is the story of three high-school-aged boys who (along with the scientist-uncle of one) build and fly the first rocket to the moon. The boys are standard popular-culture teenagers—backyard mechanics who launch model rockets and tinker endlessly with their cars. The atomic-powered *Galileo* is, for them, the ultimate science project, or perhaps the ultimate hot rod. *Salvage*, a 1979 made-for-TV movie, is the story of an enterprising scrap-metal dealer who sees money to be made in the hardware NASA abandoned during flights to the moon. Reclaiming the space junk requires a space ship, so Harry Broderick (played by Andy Griffith, in his trademark folksy style) builds his own from odds and ends in his scrap yard. The story, though technologically and financially absurd, is a rousing version of another durable American legend—the small-time businessman who does what Big Business and Big Government cannot.

The "lone inventor" model of interplanetary travel is, by a wide margin, the least plausible of the three. It is also, however, the most popular and the most consistently appealing to audiences. Popular culture exists, in part, to provide escape, and visiting other worlds in a home-built spaceship is an especially spectacular metaphor for escape.

Related Entry: Space Travel, Interstellar

FURTHER READING AND SOURCES CONSULTED

Brennan, Richard P. *Dictionary of Scientific Literacy.* John Wiley, 1992. Source of the time-and-distance figures in this entry.

Burrows, William E. *Exploring Space.* Random House, 1990. A detailed history of the robot exploration of our solar system.

Chaikin, Andrew. *A Man on the Moon.* Viking Penguin, 1994. The best of the many available histories of Project Apollo.

Ordway, Frederick I., and Randy Lieberman, eds. *Blueprint for Space: Science Fiction to Science Fact.* Smithsonian Institution, 1992. The history and future prospects of interplanetary flight; emphasizing depictions in pre-1960s popular culture.

Zubrin, Robert. *The Case for Mars: The Plan to Settle the Red Planet and Why We Must.* Free Press, 1996. A detailed explanation, for nonspecialist audiences, of how Mars could be colonized using currently available technology.

Space Travel, Interstellar

The distances between stars are so that they are measured in light-years—the distance that light, moving at 186,000 miles each second, can travel in a year. Alpha Centauri, the star closest to us, is 4.3 light years away. The other stars in our "neighborhood" of the Milky Way galaxy range from ten to twenty light-years away. The galaxy itself is roughly 100,000 light-years across, and its nearest galactic neighbor is 2,000,000 light-years away. The special theory of relativity predicts that travel at or beyond the speed of light is impossible. Crossing the vast gulfs of interstellar space thus requires a choice between two unattractive options: committing multiple human lifetimes to the trip, or finding a way around the light-speed barrier.

GENERATION SHIPS AND SUSPENDED ANIMATION

There are, in theory, two ways to send human crews to other star systems at sublight speeds. One option is to place the crew members in some form of suspended animation or cryogenic storage, to be awakened by the ship's automated systems when it reaches its destination. The advantage of this system is that the same crew that began the journey can also complete it, their dedication to the mission undiminished by tedious years (possibly preceded by the lifetimes of preceding generations) spent in transit. A second option is to build a spaceship big enough to be a self-contained, self-propelled world capable of sustaining not only the original crew but, in time, their children and grandchildren. The offspring of the original crew, grown to maturity by the time the ship reached its destination, would carry out the mission that their parents and grandparents had begun.

Both of these "slow boat" methods of interstellar travel would pose enormous technical challenges. Engines capable of driving even a modest-sized ship at even a substantial fraction of the speed of light have yet to develop beyond the conceptual stage. Cryogenic storage of human beings has yet to be attempted and may not even be possible. Creating a self-contained ecology that could supply a "generation" ship with food, water, and air for decades would require knowledge and expertise we do not yet possess. Such a ship, capable of carrying hundreds or thousands of people, would be an engineering project far beyond anything yet undertaken in space. Both methods would also pose large operational challenges. The success of a "sleeper" ship, for example, would depend on the human crew's ability to function unimpaired soon after being awakened. The success of a generation ship would, on long voyages, depend on the willingness of the "middle" generations to do their expected part though they would live and die aboard the ship without ever setting foot on a planet. Voyages using either method would necessarily be one-way trips. A return voyage would be prohibitively costly, and the travelers would "come home" to an Earth that had long since forgotten their (or their ancestors') departure.

FASTER-THAN-LIGHT TRAVEL

The technical challenges posed by faster-than-light travel to the stars are likely to be greater than any posed by the "slow boat" methods. They may well be insurmountable. The most obvious road to faster-than-light travel—the harnessing of more and more powerful engines—is apparently barred by Einstein's Special Theory of Relativity (see "Relativity"). Less obvious ones, such as exploiting the natural curvature of space and creating "shortcuts" between distant points, depend on assumptions that may be impossible to verify and technology that may be impossible to build.

Popular culture's depictions of interstellar travel portray faster-than-light travel routinely but suspended-animation and generation ships only rarely. Portrayals of all three methods gloss over the technological challenges involved, in order to focus on their sociological effects. The ships used for interstellar travel in popular culture receive second billing, far behind the people who ride them.

FASTER-THAN-LIGHT TRAVEL IN POPULAR CULTURE

Popular culture tends, by convention, to leave the technical details of faster-than-light travel obscure. Faster-than-light spaceships function, in the imagined future, much as trains, ships, and airliners do in stories about the present and recent past. They are a dramatic convenience, a means of moving characters around as quickly as possible. It is not surprising, therefore, that popular culture's depictions of faster-than-light space travel are very similar to its depictions of, especially, long-distance sea travel in our own time.

These reflections of the present and recent past in the imagined future are often extremely specific. The interstellar traders of Poul Anderson's "Polesotechnic League" stories and C.J. Cherryh's "Merchanter" stories recall the European traders who set sail in the wake of Columbus and Magellan. Anderson's hero Nicholas Van Rijn, with his ample belly and fine clothes, even looks the part of a seventeenth-century Dutch merchant prince. David Feintuch's Nicholas Seafort and A. Bertram Chandler's John Grimes both rise, over the course of many books, through the ranks of space navies remarkably similar to the seagoing ones on Earth. The refugee fleet featured in the TV series *Battlestar Galactica* (1978–1979) bears a strong resemblance to a World War II convoy, creeping through enemy territory at the speed of its slowest vessel. The *Saratoga*, the spacecraft carrier featured in the TV series *Space: Above and Beyond* (1995–1996) is modeled—down to details as small as the commodore's gold-embroidered, navy-blue baseball cap—on twentieth-century aircraft carriers. A modern-day naval aviator would feel entirely at home aboard her, at least until he looked outside.

Travel across oceans has, for centuries, been available to daring individuals as well as governments and corporations. Popular culture, perhaps for this reason, projects the same assumption onto faster-than-light travel. Han Solo's *Millennium Falcon*, featured in the *Star Wars* saga, is one famous result—the space-going equivalent of a tramp steamer. The decrepit-looking vessel ("You came here in *that*?" Princess Leia says when she first sees it; "You're braver than I thought!") has no regular route, no regular schedule, and no permanent home. Solo is his own pilot, mechanic, and business manager; he negotiates with prospective customers in spaceport bars. Like their counterparts in the interstellar navy and merchant fleets, independent operators like Solo are modeled on familiar figures from our own world—the pirate, the tramp-steamer captain, and the bush pilot. Solo himself, with his leather jacket and holstered pistol, could easily have stepped out of a floatplane deep in the wilds of Alaska. Skua September, featured in Alan Dean Foster's novel *Icerigger* (1974) and its sequels, has the name, bravado, and speech patterns ("Ho there, young feller-me-lad!") of a seagoing pirate. Jason(Robert Urich), the hero of the movie *Ice Pirates* (1984), takes the "space pirate" persona to its ultimate, tongue-in-cheek conclusion—he storms aboard his victims' spacecraft wearing high leather boots and wielding a cutlass.

SLOWER-THAN-LIGHT TRAVEL IN POPULAR CULTURE

The decades or centuries-long voyages required for interstellar travel at sublight speeds have, by contrast, no parallel in human history. Even the longest sea voyages of the sailing-ship era lasted only about five years, and most were significantly shorter. Sailors who departed on such voyages could expect, if all went well, to return home and find their home ports substantially as they had left them. One-way voyages to new homes, such as those made by immigrants, took only weeks, at most months. Stories about "slow boat" interstellar travel are virtually required, therefore, to say something about the new social and cultural structures that such travel might create. The space and effort this requires helps to explain why "slow boat" interstellar travel is rare in popular culture and, when it does appear, confined almost exclusively to novels.

Robert A. Heinlein's short novel "Universe" (1940) is the story of a generation ship whose inhabitants, after centuries in space, have forgotten that they live aboard a spaceship. Michael P. Kube-McDowell's novel *The Quiet Pools* (1992) takes place amid the imminent departure of humankind's "best and brightest" on a one-way trip to the stars. James P. Hogan's novel *Voyage from Yesteryear* (1982) depicts a human colony in the Alpha Centauri system whose members had been sent on their interstellar voyage as DNA. Raised by shipboard robots rather than human parents, they develop a society radically different from any on Earth— a sore point with Earth-born colonists who arrive later. *The Legacy of Heorot* (1987) and its sequel, *Beowulf's Children* (1995), are set on a planet whose initial human colonists are handicapped by the unexpected effects of long-term cryogenic storage on their brains. The authors—Larry Niven, Jerry Pournelle, and Steven Barnes—sketch a society torn between the cautious, group-oriented "Earth Born" generation and their risk-taking, individualistic "Star Born" offspring, who know no other home.

PIONEER 10: INTERSTELLAR TRAVEL BY DEFAULT

Practical interstellar travel, whether slower or faster than light, may be a distant and perhaps unattainable dream, but strictly speaking, the age of interstellar travel has already begun. The robot spacecraft *Pioneer 10*, launched in March 1972 on the first close fly-by mission to Jupiter, completed its assigned tasks the following year. Then, by design, it headed out of the solar system, crossing the orbit of Pluto in 1990 and heading into deep space. *Pioneer 10* is the slowest of all "slow boats" to the stars. It is, at this writing, seven billion miles from Earth—a vast distance, but only a tiny fraction of the way to the nearest stars along its path. A gold-anodized aluminum plate fastened to it proclaims, in engraved pictures, "We made this machine. This is what we look like and where we live." Whether or not it is ever found or read, that message is the payload of humankind's first attempt at interstellar space flight.

Related Entries: Relativity; Space Travel, Interplanetary; Speed of Light

Model of a *Voyager* spacecraft. *Voyager* 1 and 2 and their predecessors, *Pioneer* 10 and 11, are humankind's first interstellar spacecrafts. Traveling at nearly a million miles a day, the *Voyagers* will pass the closest stars along their routes in about 40,000 years. Courtesy of the National Space Science Data Center and Dr. Bradford A. Smith, *Voyager* team leader.

FURTHER READING

"Interstellar Travel." *Astrobiology: The Living Universe.* 27 November 2000. <http://library.thinkquest.org/C003763/index.php?page=findlife03>. Part of an extensive site; links to pages on sleeper and generation ships.

MacVey, John. *Interstellar Travel.* Scarborough House, 1991. A nontechnical survey of both sublight and faster-than-light options.

Mallove, Eugene F., and Gregory L. Matloff. *The Starflight Handbook: A Pioneer's Guide.* John Wiley, 1989. A nuts-and-bolts guide; assumes the light barrier cannot be broken.

Schmidt, Stanley, and Robert Zubrin. *Islands in the Sky: Bold New Ideas for Colonizing Space.* John Wiley, 1996. Section 4 deals with interstellar travel.

Wolverton, Mark. "The Spacecraft That Will Not Die." *American Heritage of Invention and Technology.* Winter 2001, 47–58. A compact history of the *Pioneer 10* spacecraft and its career.

Speed of Light

Light moving in a vacuum travels about 186,000 miles (or about 300 million meters) each second. Albert Einstein's Special Theory of Relativity predicts that at the speed of light, the mass of a moving object (and the energy required to move it) would become infinite. This makes the speed of light a universal speed limit that no moving object can reach, much less surpass. The existence of this "light barrier"—unlike the "sound barrier" once believed to imperil aircraft flying close to Mach 1—is predicted by a well tested and widely accepted theory. The sound barrier was first "broken" experimentally in 1947, and within a quarter-century it was being broken routinely by commercial as well as military jets. The laws of physics suggest, however, that the light barrier may be unbreakable. Cartoons, t-shirts, and bumper stickers marketed to scientists poke gentle fun at the idea, proclaiming (in imitation of 1970s public service ads for the fifty-five-mile-per-hour national speed limit): "186,000 miles per second: It's not just a good idea—it's the law."

Travel at or beyond the speed of light is central to popular culture's depictions of the future. Leaving our solar system and moving freely among the stars all but demands it. Our nearest stellar neighbor, Alpha Centauri, is 4.3 light years away—that is, roughly as far as light travels in four and a half years. Even if we could move at 90 percent of the speed of light, a round trip would still consume a decade. Voyages beyond our nearest neighbors at sublight speeds would be measured in decades, or even lifetimes, and communication would be only slightly faster. Such voyages would be functionally one-way trips. A returning ship would be greeted by the children and grandchildren of those who dispatched it.

Few stories about space travel acknowledge the realities of the light barrier and the length of interstellar voyages made at sublight speeds. Far more find ways to circumvent the light barrier and its implications. The details of faster-than-light travel are seldom spelled out in popular culture, but three basic methods stand out.

The first method is to pretend that the light barrier, or the enormous interstellar distances that make it a problem, simply do not exist. The TV series *Space 1999* (1975–1977), in which the moon drifts through space encountering a new planet each week, depends on this premise. So does the *Superman* saga, whose hero travels from Krypton to Earth in a matter of weeks. The second method is to invent shortcuts that, like the "chutes and ladders" of the classic children's board game, allow quick trips between widely separated points. Sometimes the shortcuts are natural, such as the tunnellike "wormhole" that links two distant quadrants of the galaxy in the TV series *Star Trek: Deep Space Nine* (1993–2000). More often they are artificial, such as the "jump gates" that, in the TV series *Babylon 5* (1993–1998), link distant points through a mysterious "fourth dimension."

The third, and by far the most popular, method is to invent engines that by exploiting some undisclosed loophole in the laws of physics make faster-than-light travel possible. These engines appear in many guises, from the generic "hyperdrive" used in the *Star Wars* saga to brand-name versions like the "Alderson Drive" used in Larry Niven and Jerry Pournelle's collaborative novels *The Mote in God's Eye* (1974) and *The Gripping Hand* (1993). Apart from a few exceptions designed by physicist-authors, the inner workings of these engines are left deliberately vague. They are, like matter transmitters and time machines, little more than magic wands dressed up in a thin veneer of invented jargon, like "warp core" and "dilithium crystals."

Ironically, even in stories where spaceships routinely exceed the speed of light, natural phenomena are limited by it as they are in the real world. The Puppeteers, an alien race featured in the books of Larry Niven's "Known Space" series, flee their home worlds in advance of a titanic explosion at the galaxy's core. The effects of the explosion move at the speed of light, but fortunately for the Puppeteers, their ships can move faster. The radio and television signals from Earth that alert aliens to our presence in *Contact* (novel 1985, film 1997) move at the speed of light and (accurately) take about thirty years to reach the Tau Ceti system. Ellie Arroway, the scientist chosen as Earth's first emissary to Tau Ceti, makes the same trip in a matter of minutes, with the benefit of alien technology. In many science fiction stories, it seems that the only thing that crosses interstellar space at speeds as slow as 186,000 miles per second is light itself.

Related Entries: Relativity; Space Travel, Interstellar; Speed of Sound

FURTHER READING

Dann, Jack M., and George Zebrowski. *Faster than Light*. Harper and Row, 1976.
   A fiction anthology, including some speculative nonfiction articles.

Krauss, Lawrence M. *The Physics of Star Trek*. Basic Books, 1995. Chapters 2–5 discuss faster-than-light travel in general terms.

Roman, Thomas, and Martin Bucher. "Ask the Experts: Physics." *Scientific American*. 20 December 1999. 27 November 2000. <http:// www.sciam.com/ askexpert/physics/physics57/physics57.html>. Lucid discussion of faster-than-light travel, with key terms defined.

Speed of Sound

Sound is a form of energy. It is generated when a vibrating object sets the molecules of an elastic medium (like air or water) vibrating. Humans and other animals hear sound when the vibrating medium vibrates their eardrums, which in turn transmit electromagnetic pulses to the brain. Sound cannot be transmitted without a medium, and the speed at which it travels depends on the medium's composition and density. The speed of sound in air at sea level is 1,125 feet per second, or 740 miles per hour. Speeds above that of sound are measured by Mach numbers; Mach 1 is the speed of sound, Mach 2 twice the speed of sound, and so on.

The speed of sound has, since the early 1940s, been a target for aircraft designers equivalent to the four-minute-mile in running and the 200-mile-per-hour lap in automobile racing. This image is rooted in the now-outdated idea that achieving that speed posed unique hazards to plane and pilot.

An aircraft in flight generates sound waves that propagate away from it in all directions. The behavior of the waves propagated in front of the aircraft varies greatly with the aircraft's speed. At speeds well below that of sound, they outrun the aircraft; at speeds above that of sound, the aircraft outruns them, generating a "sonic boom" as it passes observers on the ground (Brain). At speeds close to that of sound, however, the waves keep pace with the aircraft, creating a zone of turbulence that moves along with it. Aircraft designed to fly at subsonic speeds can suffer severe buffeting, loss of lift, and even loss of control when they encounter such turbulence ("Sound Barrier"). The subsonic fighter planes of the mid-1940s sometimes entered this turbulent "transsonic" realm in high-speed dives. Pilots' brief encounters with it—often terrifying, some-times fatal—helped to create the idea of a "sound barrier" that tore planes apart as they approached Mach 1.

The fictionalized, documentary-style films *Breaking the Sound Barrier* (1952) and *The Right Stuff* (1983) both portray the "sound barrier" as an

almost-physical presence—a dragon blocking the road that leads to supersonic flight. The two films' pilot-heroes set out, like modern-day
knights, to slay it. *Breaking the Sound Barrier*'s fictional Phillip Peel (John
Justin) is a stiff-upper-lipped Englishman, and *The Right Stuff*'s historical
Chuck Yeager is a laconic, drawling American, but they share a crucial
character trait—both remain calm and analytical as they enter the unknown. It is this coolness under pressure that allows them to defeat the
sound barrier. Peel discovers how to avoid loss of control, and Yeager
realizes that the turbulence he feels at Mach 1 is brief and transient.

Yeager made the real world's first supersonic flight in an experimental
X-1 rocket plane in 1947. Over the next decade, test flights showed that
aircraft designed for supersonic flight could pass through the turbulence
with ease. Military jets designed to exceed Mach 1 entered production
in the late 1950s, and their widespread use demystified supersonic flight.
Mach 1 became, like the four-minute mile in running, a performance
benchmark that divided the very fastest from the rest of the field. The
idea that supersonic aircraft were the highest of high technology made
them potent symbols of technological prowess. They played starring
roles at air shows, in military recruiting advertisements, and in films
such as *Top Gun* and *Iron Eagle* (both 1986). The public face of U.S. military aviation between the Vietnam War (1964–1973) and the Persian Gulf
War (1991) focused on the blazing speed of American combat aircraft
rather than the destructive power of the weapons they carried.

Soviet, American, and European governments funded the development of supersonic transports (SSTs) in the mid-1960s for similar image-
based reasons. Ever-increasing speed had been the hallmark of the airline
industry since the 1920s, and the step from subsonic to supersonic
seemed to government leaders the next natural step in the progression.
Airline executives and the public, however, less concerned with sym-
bolism than with practical details, disagreed. American SST projects died
on the drawing board, and the Soviet Tu-144 debuted with great fanfare
only to fade quietly away. The Concorde, an Anglo-French design that
entered service in 1969, has been dogged by high operating costs, limited
range and capacity, and public concern over noise and pollution, and
the sixteen-month grounding of the fleet after a July 2000 crash in Paris.
Critics deftly turned the SST from a symbol of technological progress
into a symbol of all that they believed to be wrong with "big technology."

Related Entry: Speed of Light

FURTHER READING AND SOURCES CONSULTED

Brain, Marshall. "What Causes a Sonic Boom?" *How Stuff Works*. 8 November
    2000. <http://www.howstuffworks.com/question73.htm>. A brief ex-
    planation of how aircraft-generated sound waves behave.

Huntington, Tom. "Encore for an SST." *Air & Space Smithsonian*. October–November 1995: 26–33. Available online at: <http://www.airspacemag.com/ASM/Mag/Index/1995/ON/esst.html>. Discusses the commercial failure of "first-generation" SSTs and plans for "second-generation" SSTs in the early twenty-first century.

Yeager, Chuck. "Breaking the Sound Barrier." *Popular Mechanics*. October 1987: 90–92+. Available online at: <http://www.popularmechanics.com/popmech/sci/9710STMIBP.html>. The first supersonic flight, described by its now-legendary pilot.

Superhumans

"Superhuman," as a concept, is easier to grasp intuitively than to define rigorously. The word "superhuman" means literally "above human," but it is hard to imagine how a line between human and superhuman could be drawn. A "human," of course, is a member of our species, *Homo sapiens sapiens*. Our species is composed, however, not of identical organisms but of similar ones. The standard for "similar"—ability to interbreed—is loose enough to encompass a great deal of variation, as the diversity of humankind shows. How different would an individual have to be, in how many ways, to qualify as "superhuman"?

The nature of evolution also frustrates any attempt to draw such a line. Species are populations of organisms whose physical characteristics cluster around a statistical mean. Evolution shifts the mean in one direction or another. If the climate cools and individuals with more hair survive in greater numbers than those with less, the population (and its statistically "average" member) will, over the course of generations, become hairier as a result. Small, isolated populations change more quickly than large, widespread populations, but most human societies are neither small nor isolated. The further biological evolution of our species (if it occurs at all) is likely to be extremely gradual. The difference between *Homo sapiens sapiens* and a hypothetical *Homo sapiens nova* ("nova" is Greek for "new") would likely be visible not in the difference between generations N and $(N + 1)$ but in the difference between generations N and $(N + 1000)$. No generation would recognize in its immediate successors differences great enough to justify the "superhuman" label.

Genetic manipulation is far more likely to produce superhumans. If it were to become both medically possible and socially acceptable to correct genetic "flaws" in children before birth, and if the means to do so were not universally available, a gap would soon emerge between the genetically modified "haves" and the unmodified "have-nots." The process would work faster than evolution could on its own and, just as important, produce "haves" whose identities were unambiguously known from the beginning. Artifice could thus, unlike nature, quickly produce a discrete group of individuals significantly different from ordinary human beings. *How* different they would have to be to qualify as "superhuman" remains an unresolved question.

SUPERHUMANS IN POPULAR CULTURE

Neither evolution nor genetic engineering have as yet produced superhumans. Society has not, therefore, been required to confront the complex social, cultural, and legal issues that their existence would raise. Superhuman characters have been prominent for decades, however, in many forms of popular culture, notably the superhero comic books that began to appear in the late 1930s and gained new popularity in the early 1960s with the work of Jack Kirby and Stan Lee. The stories told about superhumans often deal with their social and psychological complexities of their unique status. Many writers—Kirby and Lee among them—find those story elements more intriguing than straightforward save-the-world action. A distinction between superhumans created "naturally" and those created "artificially" figures prominently in those stories. Though scientifically murky, it has deep cultural significance.

"Natural" superhumans depicted in popular culture do not owe their extraordinary abilities to any human action. "Natural" superhumans do not plot to *become* superhuman, they just wind up that way. They can be products of genetic flukes, like the title character in Olaf Stapeldon's novel *Odd John* (1935) and the *X-Men* comic book series (1964–); of accidental contamination, as in *The Amazing Spiderman* (1962–), *The Incredible Hulk* (1962–) and many other comic-book series; or of the intervention of powerful alien species, as in the movie *2001: A Space Odyssey* (1968) and Spider and Jeanne Robinson's "Stardancer" novels (1979–). Natural superhumans, whatever their origins, are products of forces they cannot control or (often) understand. This makes their existence itself "natural" and therefore, in the moral universe of popular culture, "right."

"Artificial" superhumans, created by medical intervention or human-controlled selective breeding programs, are another matter. Their creators have, by definition, violated the taboo against "playing God" introduced into popular culture by Mary Shelley's *Frankenstein* (1819) and honored ever since. The creators have—again, in the moral universe of popular culture—committed the sins of egotism and arrogance. Artificial superhumans are stained by this "original sin" even if (as is often the case) they were not complicit in their own creation. The stigma attached to artificial superhumans has been strengthened, since the 1950s, by bitter memories of the early-twentieth-century eugenics movement.

Popular in Germany, Britain, and the New World, the movement advocated improving the species by promoting reproduction among the "fit" and preventing it, by isolation or sterilization, among the "unfit." The Nazis' campaign to exterminate systematically the "unfit" took the movement's principles to their logical, horrific conclusion.

The natural/artificial distinction, however arbitrary, runs deep in any discussion of superhumans. "Natural" origins suggest an orderly universe in which radical changes have (implicitly) divine sanction. "Artificial" origins raise the specter of the Nazis and the millions sacrificed to their dream of creating a "master race."

Natural Superhumans

Natural superhumans are most visible when they use their unique abilities to help normal folk, but the most memorable stories about them tend to be character rather than action driven. Characters who are born superhuman routinely find themselves searching, in such stories, for acceptance in a world that neither trusts nor understands them. In J.K. Rowling's best-selling fantasy novels, Harry Potter and his fellow wizards go to great lengths to conceal their powers from nonmagical folk. Like the mutant "X Men," they can be truly themselves only in the company of others like them. Stapledon's *Odd John* is a tragedy, because its hero and his fellow "supernormals" *cannot* find such sanctuary in a world of hate and fear-filled "normals."

Characters who become superhuman by accident must come to grips with a body that has, suddenly and without warning, undergone changes far more radical than those brought by adolescence or old age. Those changes permanently alter the routines of daily life; Dr. Bruce Banner must constantly control his anger, lest it trigger a transformation into the Incredible Hulk. They also bring new responsibilities; the superhuman powers that make Peter Parker into Spiderman create, for him, a moral obligation to *use* them for Good. Superhuman powers, a boon in action-oriented stories, are a burden in character-driven stories. Superhumans become sympathetic, in such stories, because what is (relatively) easy for ordinary folk is desperately difficult for them.

The routine depiction of "natural" superhumans as socially isolated means that they can be sympathetic, to a degree, even when they behave badly. Magneto, chief villain in the movie *X-Men* (2000), lost his parents to the Nazi death camps as a boy and, as an adult, fears a new Holocaust at the hands of fearful "normals." Dr. Charles Xavier—Magneto's friend, contemporary, and fellow mutant—confronts him, not in order to destroy him but to heal his psychological wounds and restore his faith in "normals." Ender Wiggin, the boy-hero of Orson Scott Card's novel *Ender's Game* (1985), exhibits superhuman intelligence and as a result becomes an unwitting pawn in a genocidal war against intelligent aliens. Stephen King's *Carrie* (1974) is tormented by high school classmates who deride her as "different" but do not realize that she has latent telekinetic powers. Driven over the brink by a spectacular act of public humiliation, she lashes out at them in an equally spectacular act of telekinetic vengeance.

Artificial Superhumans

Artificial superhumans, if they consent to or approve of the process that created them, carry the same moral taint as their creators. Khan Noonian Singh, a product of the "Eugenics Wars" who escapes from exile twice in the course of the *Star Trek* saga, is arrogant and egotistical and he has an insatiable hunger for the power that he believes is his birthright. The biologically enhanced athletes in Larry Niven and Stephen Barnes's novel *Achilles' Choice* (1991) and films like *Goldengirl* (1979) and *Rocky IV* (1985) are at least implicitly criticized both for "cheating" and for surrendering too much of their humanity in their quest for fame and glory. Charly Gordon, the mentally retarded janitor who narrates Daniel Keyes's "Flowers for Algernon" (short story 1959, novel 1966), is an especially striking example. Appealing and sympathetic at the beginning of the story, he becomes arrogant and unpleasant after an experimental medical procedure grants him superhuman intelligence. He becomes sympathetic again only when humbled by the realization that the procedure's effects are temporary.

Artificial superhumans who were *not* complicit in their own creation carry no moral taint but suffer from other problems. They are seldom well balanced or wholly at peace with themselves but rather are scarred by the knowledge of their calculated, "unnatural" origins. Lyta Alexander (Patricia Tallman), a recurring character on the TV series *Babylon 5* (1994–1999), feels intense bitterness over the knowledge that she and her fellow telepaths are products of a secret alien breeding program. Dr. Julian Bashir (Alexander Siddig), one of the principal characters on TV's *Star Trek: Deep Space Nine* (1993–2000), feels that a deep but narrow gulf separates him from his friends, because his parents had him genetically "enhanced" before birth. Neither character can escape the feeling of being a by-product of someone else's experiments, however well intentioned those experiments were.

Fictional superhumans are consistently presented as ambiguous figures. Their exploits are thrilling but also troubling, because they highlight the superhumans' effortless superiority over ordinary folk. There is every reason to believe that real superhumans—if and when they come into existence—will be regarded in much the same way.

Related Entries: Evolution; Evolution, Human; Genetic Engineering; Mutations

FURTHER READING AND SOURCES CONSULTED

Bostrom, Nick, et al. "Transhumanist FAQ." 13 May 1999. World Transhumanist Association. 7 December 2001. <http://www.transhumanist.org>. Introduction to a movement dedicated to the use of science and technology to accelerate and guide human evolution, in effect producing superhumans.
Kevles, Daniel J. *In the Name of Eugenics*. University of California Press, 1984. History of the eugenics movement in Britain and the United States.
Niven, Larry. "Man of Steel, Woman of Kleenex." In *All the Myriad Ways* (Ballantine Books, 1971). Also available at: <http://www.blueneptune.com/svw/superman.html>. Mock-serious analysis of the complexities that procreation and childbirth would present for a human/superhuman couple.
Sanderson, Peter. *Marvel Universe*. Abradale Press, 1996. A survey of Marvel Comics' (mostly) superhuman characters.

Time Travel

We think of physical objects as having three spatial dimensions: length, width, and height. We also think of them as having measurable life-spans. Time is, therefore, often described as "the fourth dimension" (the fourth *spatial* dimension posited by physicists and mathematicians is well outside the realm of everyday experience). We can move freely through all three spatial dimensions. We can also transmit information through space at times, and to places, of our own choosing. Our ability to move ourselves or to transmit information through the fourth dimension, time, is far more limited. The past is closed to us, and the future open only in the narrowest and most limited sense.

TIME TRAVEL TODAY

Time travel into the future is, in one sense, routine. It happens to everyone, whether they want it to or not. A majority of the individuals born in the United States this year will live to see the (very different) United States of fifty years from now—by living through the intervening years and aging half-a-century in the process. The often-expressed sentiment "I hope I live to see [some event]" is, in effect, a wish to time-travel in this way to a specific destination in the future. Time capsules, filled and then sealed until a specified date in the future, operate on the same principle. Individuals awakened from long comas or (as the result of a medical breakthrough not yet accomplished) from cryogenic storage "time travel" in the same way but experience it differently. They still age (though perhaps more slowly), but they are not aware of the passage of time.

None of these methods, however, is "time travel" in the usual sense of the term. None of them offers instantaneous access to the future, none of them offers any access to the past, and none of them is reversible. They are to fictional time travel what riding an inflated inner tube down

a river is to boating: a pale and unsatisfying substitute. For the foreseeable future, however, they are the best that we can hope to do.

TIME TRAVEL SOMEDAY (MAYBE)

Time travel may not be possible, because neither the past nor the future may exist as a "place" that can be visited and revisited. Consider
the difference between watching a live performance and a recording of
it. Both versions of the performance are identical, and each can be divided into a series of moments that occur one after the other. People
observing the live performance, however, can experience each moment
only once, fleetingly, as it happens in their present. People observing the
recorded performance have more control over how they experience it;
for them, each moment exists independently and can (with the proper
equipment) be revisited over and over. Both analogies have their defenders, but only the second permits time travel.

Many physicists—Albert Einstein and Stephen Hawking among
them—have argued that time travel *is* theoretically possible. Einstein's
General Theory of Relativity treats space and time as aspects of a single
entity, called "space-time," that becomes distorted by the presence of
massive objects and at speeds at or near that of light. Travel through
time should be possible, therefore, given the right equipment.

Designing and building "the right equipment" will be a significant,
perhaps insurmountable, challenge, however. The most widely accepted
method of time travel to the future requires that the time traveler move
at speeds close to that of light. The passage of time will, according to
Einstein's theory of special relativity, slow to a crawl within the fast-
moving "universe" of the time-traveler's vehicle. She will experience the
trip as the passage of a few days, while years pass on Earth. Returning,
she will find herself in the future. Returning to her own time, according
to some theorists at least, will require her to accelerate her vehicle *past*
the speed of light, reversing the flow of time and allowing her to go
home. Other proposed mechanisms for time travel include "wormholes"
that connect nonadjacent regions of space-time, and captive black holes
modified for the purpose. All of these methods would pose enormous
technological and practical problems. They would, for example, consume
energy on a scale undreamed of today.

One recent (relatively) low-energy method, proposed by physicist Ronald Mallett, involves a machine that generates two beams of light moving
in opposite directions around circular paths at the very slow speeds that
recent experiments by others have shown to be possible. The moving
light beams would, in Mallett's theory, use a modest amount of energy
but create major distortions of space and time within the circle. Someone
inside the circle would, in theory, be able to travel backward through

time simply by walking across the floor. The time traveler would, however, be able to travel back no farther than the moment at which the circle of light was generated.

A second objection to time travel, especially into the past, is metaphysical. Unless the universe is entirely deterministic, there are at any given moment many possible futures—on the morning of 22 November 1963, for example, ones in which John F. Kennedy died and others in which he lived. Travel into the past carries the risk of altering a crucial event in such a way that subsequent history is irrevocably changed. One classic paradox has a time traveler murdering his grandmother, precluding the birth of his parent and presumably his own existence.

One possible solution to this paradox is to postulate that the past is immutable and that even the most determined efforts to change it will fail. Another lies in the "many worlds interpretation" of quantum mechanics, which proposes that any event with more than one possible outcome splits the universe in which it happens into multiple universes, one in which each possible outcome takes place. A time traveler who succeeds in murdering his grandmother does not wink out of existence, because his birth occurred in a universe in which the murder failed.

TIME TRAVEL IN POPULAR CULTURE

Time travel is both theoretically possible and practically straightforward in popular culture. It may be accomplished with roomfuls of equipment and millions of dollars, as in the TV series *Seven Days* (1999–) and Michael Crichton's novel *Timeline* (1999), but is not beyond the reach of an ambitious inventor with a well-equipped workshop, as in H.G. Wells's novel *The Time Machine* (1895) and the *Back to the Future* movies (1985–1990). The heroes of Jack Finney's novel *Time and Again* (1970) and Richard Matheson's *Bid Time Return* (1975; filmed as *Somewhere in Time*, 1980) will themselves back in time. Time travel also happens, with surprising frequency, by accident. Lightning is responsible in L. Sprague DeCamp's classic novel *Lest Darkness Fall* (1949), mysterious storms in the 1980 movie *The Final Countdown*, and a blow to the head in Mark Twain's *A Connecticut Yankee in King Arthur's Court* (1889).

Instantaneous time travel in popular culture nearly always takes the traveler into the past. Time travel into the future is far less common; when it does occur, it is usually accomplished by "conventional" methods, like suspended animation. Comic-book hero Buck Rogers, for example, is overcome by a mysterious gas in the 1930s and awakens in 2419. The hero of *The Time Machine*, who uses the titular machine to cruise into the distant future, is a rare exception to this pattern.

Having sidestepped or ignored all other potential problems related to time travel into the past, popular culture embraces the possibility of al-

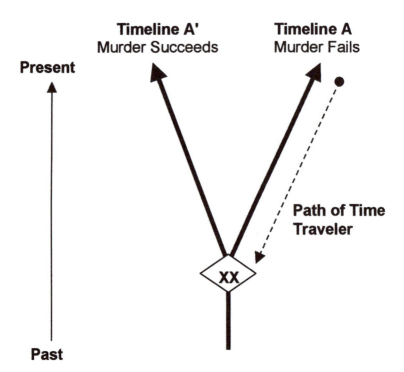

Timeline A'   **Timeline A**
Murder Succeeds   Murder Fails

Present

**Path of Time
Traveler**

XX

Past

The "Grandmother Paradox" resolved. Quantum mechanics suggests that a time traveler who attempts to murder his maternal grand-mother (at XX) creates two timelines. In timeline A he fails, ensuring that he will be born, build the time machine, and return to attempt it. In timeline A' he succeeds, precluding his mother's birth and thus his own.

tering the future. An entire subgenre of time travel stories concerns (sympathetic) attempts to rewrite history, such as Hank Morgan's efforts to start the Industrial Revolution early in *Connecticut Yankee*. A second subgenre—Poul Anderson's "Time Patrol" and Robert Aspirin's "Time Scout" stories—follow the heroes' efforts to *keep* time travelers from altering history. A third subgenre derives its dramatic tension from the problems, both ethical and practical, of playing with the fabric of the past. *The Final Countdown* asks whether the commander of a modern aircraft carrier, transported to the waters off Hawaii in early December 1941, should use jet-age firepower to crush the impending Japanese attack on Pearl Harbor. Orson Scott Card's novel *Pastwatch: The Redemption of Christopher Columbus* (1995) cautions that in changing history, the cure might be worse than the disease. Ray Bradbury's short story "The Sound of Thunder" (1952) demonstrates in its devastating final paragraphs that even seemingly trivial alterations of the past, casually or unknowingly made, could have catastrophic effects on the future.

Stories about time travel also embrace the possibility of logical paradoxes. P. Schuyler Miller's short story "As Never Was" (1942) concerns the inexplicable origins of a mysterious artifact. Found in the ruins of a museum by time travelers to the future, it is examined by present-day scientists, who take a sample from it, leaving an indelible mark. It is then placed in the museum in the ruins of which time travelers will one day discover it—unmarked. Robert A. Heinlein's stories "By His Bootstraps" (1941) and "—All You Zombies" (1959) are tours de force in which all the characters are manifestations of the same time-travelling individual. The *Back to the Future* and *Terminator* (1984, 1991) movies also make effective (and underappreciated) use of similar paradoxes.

Stories about time travel are similar in many ways to stories about faster-than-light travel through space. Both require rejecting, sidestepping, or ignoring much of what we believe to be true about the natural world. Both have been buoyed by suggestions they *might* be possible in the future. Both, finally, open up a vast range of dramatic possibilities: allowing characters to go elsewhere (or elsewhen) in ways undreamed of in the real world.

Related Entries: Alternate Worlds; Light, Speed of; Relativity

FURTHER READING AND SOURCES CONSULTED

Brooks, Michael. "Time Twister." *New Scientist*. 19 May 2001. Detailed report on Ronald Mallett's theory of time travel by slow light.

Gott, J. Richard, III. *Time Travel in Einstein's Universe: The Physical Possibilities of Travel through Time*. Houghton Mifflin, 2001. A comprehensive discussion by a leading researcher in the field.

Nahin, Paul J. *Time Machines: Time Travel in Physics, Metaphysics, and Science Fiction*. Springer-Verlag, 1993. Time travel in science, philosophy, and popular culture.

Preston, Steve. *Time Travel*. 13 June 2001. <http://freespace.virgin.net/steve.preston/Time.html>. A superbly organized, clearly written site with explanations for novices and experienced physics students.

Time Travel Research Center. 12 June 2001. The Time Travel Research Center. 13 June 2001. <http://www.time-travel.com>. An enormous collection of time-travel information: physics, metaphysics, mysticism, art, literature, and film.

UFOs

The acronym "UFO" stands for "unidentified flying object." No one disputes that they exist or that the majority of UFOs sighted turn out to be airplanes, balloons, or atmospheric phenomena. The controversy involves the remainder—those that do not fall, easily and obviously, into one of those categories. UFO skeptics argue that the remaining sightings can also be explained in familiar scientific terms. UFO believers argue that the remaining sightings can best be explained as evidence of advanced alien species visiting the earth.

The skeptics' position is based on Occam's Razor, the principle of logic that, all other things being equal, simple explanations that refer to known principles and entities are more plausible than complex explanations that invoke new ones. UFO skeptics argue on those grounds that explaining anomalous sightings in the same terms as the rest is far more plausible than inventing an advanced alien civilization to account for them. Skeptics often elaborate their case by raising more specific plausibility issues. How would the hypothetical aliens cross interstellar distances, when the known laws of physics suggest that faster-than-light travel is impossible? Why would an alien species advanced enough to travel between stars be interested in humans, who can barely leave their planet? Why, if they did not wish to be detected, would such aliens not use their advanced technology to observe us from a safe distance? Why, if they sought contact, do reported alien-human encounters typically involve individuals in remote areas?

The believers' position rests on a large collection of observations that they find individually compelling and collectively persuasive. They argue that some UFO sightings stubbornly defy conventional explanations—lights in the sky, for example, that move at speeds and with maneuverability that no known aircraft could match. They recount the stories of individuals who believe that they saw evidence of, or were abducted by, alien visitors. Some members of the believers' community connect alien UFOs to cattle mutilations, crop circles, and other events they deem inexplicable. Many of them believe that the U.S. federal government knows that some UFOs are of alien origin and has the evidence—corpses and hardware from a crashed alien spacecraft—hidden at a secret military base in the Nevada desert. The government is keeping the alien origin of UFOs a secret, according to this theory, because it fears panic if the information became widely known.

Popular culture sides squarely with the believers. It has spread their views widely enough that "UFO" is now synonymous, in everyday usage, with "alien spacecraft in the vicinity of Earth." The strange lights that characters notice in the sky are virtually never anything as pedestrian as a passing airplane, a stray weather balloon, or moonlight reflecting off a cloud. They are spacecraft on their way to invade us, as in *War of the Worlds* (novel 1899, radio play 1938, film 1953); make friends with us, as in *Close Encounters of the Third Kind* (film, 1977); or observe us secretly, as in *Third Rock from the Sun* (TV series, 1995–2001).

The more elaborate (or, from the skeptic's perspective, more outrageous) views of believers are well represented in popular culture. TV series from *My Favorite Martian* (1963–1966) and *The Invaders* (1967–1968) to *Roswell* (2000–2001) have been predicated on the idea that aliens already live among us, unknown to most. Sensationalized "documentaries" like the movie *Hangar 18* (1980) and the Fox Network TV special *Alien Autopsy* (1995; declared a hoax by the network in 1998) promise to expose the government's role in hiding physical evidence of alien UFOs. The long-running series *The X-Files* (1993–) uses alien abductions and a government actively complicit with aliens as part of its underlying "mythology"—the hidden background story gradually revealed over the course of the series. Novelist Whitley Streiber has written two bestselling works, *Communion* (1987) and *Transformation* (1988), about his purported encounters with aliens. Art Bell gives nationwide radio airplay to similar claims on his syndicated late-night call-in show. "Aliens" with slight bodies, gray or green skin, and huge, slanted black eyes are now ubiquitous popular culture icons. Roswell, New Mexico, has built a booming tourist industry around a famous 1947 UFO sighting. It is now the site of the International UFO Museum and Research Center, and the subject of countless documentaries like the History Channel's *Roswell: Secrets Revealed* (2000).

Commentators offer many explanations for the popularity and durability of the belief that some UFOs are alien spacecraft: fear of nuclear war or foreign invasion, a need to reinterpret old myths, a burgeoning distrust of government. One factor clearly at work, however, is fun. Popular culture, whatever its other functions, is also entertainment, and it is far more entertaining to believe that alien spacecraft are visiting us than that lights in the sky are nothing more than the navigation lights of an Atlanta-bound airliner, refracted through passing clouds.

Related Entries: Life, Extraterrestrial; Space Travel, Interplanetary; Space Travel, Interstellar

FURTHER READING AND SOURCES CONSULTED

Alschuler, William R.. *The Science of UFOs*. St. Martin's, 2001. A detailed brief for the skeptics' position.
Hynek, J. Allen. *The UFO Experience*. Marlowe, 1999. A detailed brief for the believers' position.
Sturrock, Peter A., and Laurance A. Rockefeller. *The UFO Enigma: A New Review of the Physical Evidence*. Aspect, 2000. Detailed scientific analysis of reported UFO sightings. Treats the issue as unresolved.

Vacuum

A vacuum is a region of space containing virtually no matter. Vacuums exist naturally in space, and they can be created artificially by removing the air from inside a sealed container. Neither form of vacuum can be "perfect," or totally devoid of matter—space has no boundaries, and neither vacuum pumps nor gas-tight seals are 100 percent efficient—but in both cases the difference is negligible. The pressure exerted by surrounding gas molecules on an object 800 kilometers above the Earth is roughly a trillionth of that exerted at the earth's surface. Vacuums created in laboratories and those occurring naturally in interstellar space have even fewer molecules and lower pressures.

The nearly total absence of matter in a vacuum makes it an environment unlike any naturally occurring on Earth. It cannot support combustion, since it contains virtually no oxygen. It cannot transmit heat or sound; the molecules it contains are too widely scattered. It exerts virtually no drag on objects passing through it and has virtually no filtering effect on ultraviolet radiation from the sun.

Most important, from a space traveler's perspective, a vacuum is utterly hostile to human life. An astronaut entering space unprotected would be instantly unable to breathe. The oxygen already in his lungs and bloodstream would be enough to supply his body's needs for about fifteen more seconds; then, as deoxygenated blood began to move through his arteries to his brain and other organs, he would begin to lose consciousness. Death from oxygen starvation would follow in a matter of minutes. The dying astronaut would be aware of the water on his tongue vaporizing into space. He might, if something prevented the pressure in his ears from equalizing, suffer ruptured eardrums. His extremities would swell somewhat, but his body would remain intact, the strength of its skin limiting the expansion of the fluids inside.

Stories set in outer space generally use vacuum the way sea stories use the water—as a hostile environment that tests the main characters' endurance and ingenuity. The most famous of these man-against-vacuum scenes occurs midway through the film *2001: A Space Odyssey* (1968). Astronaut Dave Bowman, locked out of the spaceship *Discovery* by its crazed onboard computer, must enter space without a helmet in order to reach and operate the airlock that will allow him to breathe again. His nerve-wracking "swim" through an airless corridor, filmed in total silence, recalls many equally desperate fictional journeys through the compartments of partially flooded ships and submarines.

Dave Bowman does not explode on his way to the safety of the airlock, a fact that reflects the scrupulous attention paid to the scientific details of the story by director Stanley Kubrick and scenarist Arthur C. Clarke. Audiences, however, are often surprised; people exposed to vacuum *do* explode, after all, in films such as *Outland* (1981) and *Total Recall* (1990). The popularity of body-exploding-in-vacuum scenes is easy to understand. Commercial movies like *Outland* are shaped, in ways that print fiction and more experimental films like *2001* are not, by the need to tell a story using images that are both striking and readily comprehensible to viewers. A body exploding in a vacuum suggests "death" more clearly and strikingly than could a body succumbing to oxygen starvation. It is unambiguous, even when filmed at a distance to downplay the gore. Finally, it provides the victims—nearly always Bad Guys—with a satisfyingly unpleasant fate. Scientific accuracy is, for most filmmakers, a small sacrifice for such dramatic returns.

The same preference for visual interest over scientific precision also drives other vacuum-related movie conventions. Exploding spaceships and space stations create enormous fireballs in the absence of any oxygen and generate thunderous booms that no observer should be able to hear. Spacecraft swoop and bank in ways that would make sense only if they were maneuvering through an atmosphere and deflecting the flow of air over their wings, as airplanes do, in order to change direction. The climactic attack by rebel fighters on the Death Star in *Star Wars* (1977) could just as well be the climactic attack by American dive bombers on the Japanese carrier fleet in *Midway* (1976). Telling the story in familiar visual terms—banking fighters and loud, fiery explosions—saves filmmakers from the demanding work of thinking through the scientific "ground rules" of an alien environment like outer space and explaining them to the audience.

Thinking through those implications can pay benefits, however. The memorably ominous ad line used to promote *Alien* (1979) reflects a fundamental truth about vacuum: "In space, no one can hear you scream."

Related Entry: Space Travel, Interplanetary

FURTHER READING AND SOURCES CONSULTED

Landis, Geoffrey. "Explosive Decompression and Vacuum Exposure." 8 January 2001. 21 August 2001. <http://www.sff.net/people/Geoffrey.Landis/vacuum.html>. A semitechnical explanation, with references to medical literature.

NASA Goddard Space Flight Center. "Ask a High Energy Astronomer: Human Body in a Vacuum." 21 August 2001. <http://imagine.gsfc.nasa.gov/docs/ask_astro/answers/970603.html>. A nontechnical explanation, used in preparing this entry.

Venus

The more we learn about Venus, the less hospitable it looks. Seen from Earth with the naked eye or a low-power telescope, it is a bright, featureless disc, blazing brilliant white with reflected sunlight. Seen through the powerful telescopes available to nineteenth- and twentieth-century astronomers, it is a swirling mass of impenetrable clouds. Seen from beneath the cloud layer, as it was by Soviet spacecraft beginning in the mid-1970s, it looks like artists' depictions of hell. The cloud layers traps much of the heat that the sun pours into Venus, keeping the planet's surface hot enough (at 500 degrees Celsius) to melt lead. The dense atmosphere exerts a hundred times more pressure than Earth's does, pressing down with a more than a half-ton of force on every square inch of the surface. Sulfur dioxide, spewed into the atmosphere by Venus's active volcanoes, reacts with atmospheric gasses and falls as a rain of sulfuric acid. As one of the few geologically active bodies in the solar system (Earth and a handful of large moons are the others), Venus is a fascinating world to study. Unfortunately, it is also a difficult world to study; not even robots can survive long on its nightmarish surface.

Venus, as an object in the sky, has been part of popular culture since antiquity. The sunlight reflected from the tops of its clouds makes it one of the brightest objects in the night sky and one of the easiest to recognize. Its brightness is even more striking because it is only visible in the hours just after sunset or just before sunrise, when distant stars appear muted against the sky, not yet fully dark. Venus's brilliant whiteness almost certainly accounts for its being named after the Roman goddess of love and its use as a symbol of traditional "womanly" virtues: purity, chastity, and austere beauty. It even suggests the gleaming white marble used, by classical sculptors, to model idealized versions of the female body. John Gray's *Men Are from Mars, Women Are from Venus* (1992) does the same, using "Venus" as shorthand for women's psychological makeup and "Mars" for men's.

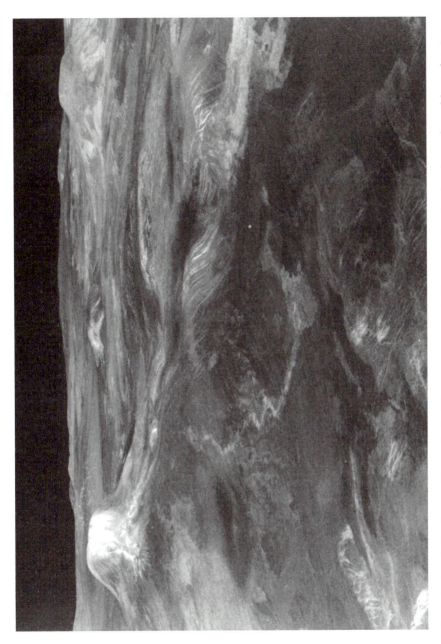

Venus beneath the clouds. The surface of Venus, reconstructed here using radar, has been shaped by volcanism. The plains in the foreground are lava flows; the volcanoes in the background are 6,300-foot-high Sif Mons (right) and 10,000-foot-high Gula Mons (left). Courtesy of the National Space Science Data Center and Dr. Gordon H. Pettengill, Magellan Experiment Principal Investigator, the *Magellan* project.

The symbolic link between Venus and Woman also appears, in a far less abstract form, in low-budget science fiction movies. When the comic team of Bud Abbott and Lou Costello land on Venus in the misleadingly titled film *Abbott and Costello Go to Mars* (1953), they find it occupied by an all-female society whose members are played by beauty-contest winners. *Voyage to the Planet of Prehistoric Women* (1968) introduces a similar society, its members clad in seashell bikinis and led by the bleached-blonde, large-breasted Queen Moana. *Vampire Vixens from Venus* (1996) banished any lingering shreds of doubt about the sexual undertones of women-from-Venus films. The three title characters, actually disguised alien drug dealers, seduce unsuspecting Earth men in order to drain (fatally) their life-energy from them at the peak of their sexual arousal.

The most enduring images of Venus in popular culture, though, have less to do with women than with weather. Scores of writers, movie directors, and comic-book artists, taking their cue from the swirling clouds, have pictured it as a world of swamps, lush vegetation, and perpetual dampness. These fictional versions of Venus are reminiscent of the impenetrable African jungles of Joseph Conrad's *Heart of Darkness*, and they are equally hostile to humans. Robert A. Heinlein used such a background for his short story "Coventry" (1941), imagining Venus as a dumping ground for troublemakers exiled from Earth. In "The Green Hills of Earth" (1947), Heinlein's ballad-singing main character sums the planet up this way: "We rot in the swamps of Venus, we retch at her tainted breath/Foul are her flooded jungles, swarming with unclean death." Ray Bradbury, better known for stories set on Mars, sketched a Venus as grim as Heinlein's in his story "All Summer in a Day" (1954). Like a nightmare vision of Seattle conjured up by a lifelong resident of Arizona, Bradbury's Venus is always rainy. The rain falls all day, every day, pounding the vegetation flat and filling the air with a smothering dampness. The human colonists who live there are resigned to it, their children unaware that the rain *could* ever stop. The story's pivotal event—a single hour when the rain stops and sun shines—is like a biblical miracle, both because it seems to suspend the laws of nature and because it occurs only once every seven years.

Neither image of Venus survived contact with reality; intense heat, crushing atmosphere, and corrosive rain were too different from beautiful women and swampy jungles ever to be reconciled with them. Since 1980, Venus has all but disappeared from popular culture. When John Gray writes that "Women Are from Venus," he is invoking what the ancient Romans saw—a bright, pure white light in the evening sky.

Related Entries: Mars; Moon

FURTHER READING

Arnett, Bill. "The Nine Planets: Venus." 27 April 1999. 21 August 2001. <http://seds.lpl.arizona.edu/nineplanets/nineplanets/venus.html>. Overview of current knowledge, with pictures and radar maps.

Cooper, Henry S.F. *The Evening Star: Venus Observed*. Farrar, Straus, and Giroux, 1993. Journalistic account of the *Magellan* space probe's mapping of Venus.

Heinlein, Robert A. "The Green Hills of Earth." 1949. In *The Past through Tomorrow*, New York: Putnam's, 1967. Source of the (fictional) song lyrics quoted in this entry; "Coventry" appears in the same volume.

Marov, Mikhail, et al. *The Planet Venus*. Yale University Press, 1998. A comprehensive resource, written by and for planetary scientists.

Volcanoes

Volcanic eruptions and earthquakes both release vast quantities of energy into small areas over short periods of time, with potentially catastrophic results for humans in the vicinity. Americans' view of volcanoes is very different, however, from their view of earthquakes. Earthquakes are common enough in California that most adult Americans have seen live news coverage of at least one major one. They are also part of local history and lore in states as diverse as Alaska (Anchorage, 1964), Missouri (New Madrid, 1811), and South Carolina (Charleston, 1755). Volcanic eruptions are much fewer and farther between in the United States. They are also, except for the sudden eruption of Washington's Mount St. Helens in 1980, restricted to the outlying states of Alaska and Hawaii. Popular culture's depictions of volcanoes reflect both their potential for sudden destruction and their exotic strangeness.

The earth's solid crust, which "floats" on a sea of molten rock, is not uniformly solid. It is shot through with cracks, fissures, and chambers into which the molten rock (called "magma") can penetrate. Volcanoes are formed when molten rock, now called "lava," reaches and flows onto the surface. Volcanoes' characteristic cone shape is created by repeated eruptions that layer new lava and ash over old; the consistency of the lava and the proportions of lava and ash determine the angle of the cone's sides. Lava and ash routinely destroy property when volcanoes erupt near human settlements, burning, burying, or entombing it. Their damage is generally localized, however, and residents willing and able to move quickly can usually flee. The rapid burial of the Roman resort city of Pompeii in 79 A.D. is the exception rather than the rule. In at least one instance (the island of Heimay, off the coast of Iceland, in 1973), local residents organized a counterattack that diverted flowing lava away from an economically vital harbor.

Volcanoes do not always do their damage at a slow and stately pace, however. They can also explode, doing catastrophic damage in moments.

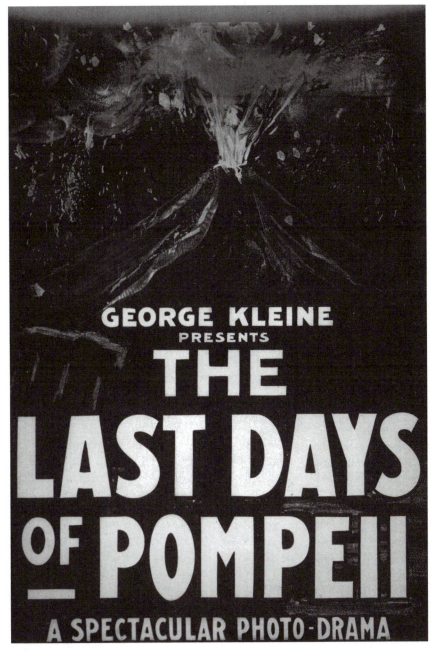

The Last Days of Pompeii. This poster for the 1913 film adaptation of Edward Bulwer-Lytton's novel uses the standard image of an erupting volcano. Fountains of fire and streams of lava replace the thick clouds of ash that actually buried the ancient Roman resort city in 79 A.D. Courtesy of the Library of Congress.

Volcanic explosions virtually destroyed the Greek island of Thera in the 1300s B.C. and the Indonesian island of Krakatoa in 1883. The 1902 explosion of Mount Pelée, on Martinique, unleashed an avalanche of superheated gas, dust, and rock that plunged downhill at sixty miles per hour and engulfed the nearby town of St. Pierre. Within moments, all but two of St. Pierre's 28,000 inhabitants were dead.

Popular culture presents volcanoes in two distinct ways. The more common, but less obvious, uses volcanoes as visual shorthand for places that (like the volcanoes themselves) are exotic and possibly dangerous. Distant volcanoes gushing lava or emitting slender, ominous plumes of smoke have become a visual cliche in depictions of prehistoric Earth. Experienced viewers of such images (cartoon or painting, still or moving) might well conclude that dinosaurs were compelled, by some ecological quirk, to live within sight of a volcano. Volcanoes, again emitting ominous plumes of smoke, also appear prominently in standard images of tropical islands. Movies from *King Kong* (1935) to *Joe Versus the Volcano* (1990) and TV series such as *Gilligan's Island* (1964–1967), *The Swiss Family Robinson* (1975–1976), and *Tales of the Gold Monkey* (1982–1983) all make use of it.

Less often, but in far more spectacular fashion, volcanoes appear in popular culture as an active threat to the main characters. Mount Vesuvius looms over Roman townspeople in *The Last Days of Pompeii* (novel 1834; films 1913, 1935, 1984). The imminent explosion of Krakatoa threatens a lost balloonist and the extraordinary island society that rescues him in William Pené Du Bois's acclaimed juvenile novel *The Twenty-One Balloons* (1947). It also threatens a far-less-interesting group of characters in the 1969 movie *Krakatoa, East of Java* (retitled *Volcano* after the studio discovered that Krakatoa was in fact *west* of Java). Movies such as *The Devil at Four o'Clock* (1961), *Dante's Peak* (1997), and *Volcano* (1997) deal with unexpected volcanic eruptions that threaten innocent people. The volcanoes and their lava flows become large and fierce, if slow-moving, monsters from which the heroes must rescue those innocents.

The 1997 film *Volcano*, set in downtown Los Angeles and advertised with the slogan "The Coast Is Toast," is notable for its climax. Its heroes make a stand against the lava, as the Icelanders did at Heimay, by cooling the front of the advancing flow with water in order to form a barrier that will divert the flow from vital parts of the city. One of the characters even makes a passing reference to Heimay as a precedent—art imitating life more precisely than is usual in popular culture's treatment of science.

Related Entries: Comets; Earthquakes; Lightning; Meteorites

FURTHER READING AND SOURCES CONSULTED

Decker, Robert, and Barbara Decker. *Volcanoes*. 3rd ed. Freeman, 1997. A comprehensive overview; the best single volume on the subject.

Fisher, Richard, et al. *Volcanoes: Crucibles of Change*. Princeton University Press, 1998. A good alternative to Decker and Decker; slightly less comprehensive on scientific issues, slightly more so on history.

McPhee, John. *The Control of Nature*. Farrar, Straus, and Giroux, 1989. Three essays, one dealing with Heimay's battle against advancing lava.

Whales

Whales, present in all the world's oceans, are air-breathing marine mammals ranging up to 100 feet in length and 150 tons in weight. They are among the most fascinating of all wild animals and among the least completely understood. The details of whale anatomy and physiology are familiar—whales have been hunted, and therefore butchered, for centuries—but the details of their mental and social lives are obscure. It is clear that whales are gregarious, but it is less clear how individual whales interact with each other. It is well known that whales are intelligent, but the dimensions of that intelligence are uncertain. It is widely believed that whales communicate using sound, but the nature and purpose of that communication are understood vaguely at best. These qualities, however imperfectly grasped, are key to the public's fascination with whales. Popular culture, in depicting whales, routinely substitutes expansive certainties for scientists' limited hypotheses about their thought, behavior, and intelligence.

Whales, as a group, figure prominently in the rhetoric of the modern environmental movement. "Save the Whales," coined in the 1970s, is still the best-known pro-environment slogan in English. (One measure of its popularity is the collection of counterslogans that play off of it: "Nuke the Whales," and, more subtly, "Save the Whales—Collect the Whole Set!") Whales are powerful symbols for many reasons; they are large, graceful, attractive mammals directly threatened by humans in a readily comprehensible way (hunting). The key to their symbolic power, however, lies in the parts of their lives that we know the least about—their intelligence, social behavior, and communications. The modern environmental movement, born in the 1960s, has traditionally called for both a deeper understanding of nature and a more thoughtful relationship with it. Whales starkly illuminate both needs. It would, environmentalists argue, be the height of narrow-minded arrogance to drive an intelligent species to extinction while remaining ignorant of what would be lost as a result.

Individual whales are far less common in popular culture than are generic whales. They are also far less common than individual dolphins, and they are usually depicted in very different terms. Fictional dolphins tend to behave like faithful dogs, with humans squarely at the center of their mental universes; the humans that befriend these dolphins treat them, in turn, as pets rather than as equals. Fictional whales, on the other hand, have far more autonomy. They are as intelligent as the humans they encounter and, implicitly, able to feel similar emotions. When they interact with humans, they do so as equals. Indeed, they often behave in ways that would not seem out of place in human characters.

The legendary white sperm whale in Herman Melville's *Moby-Dick* (1856) is, for example, a worthy opponent for his pursuer Captain Ahab. His tenacity and cunning matches Ahab's, and (though it may not have been Melville's intention) his behavior suggests a personal vendetta against the one-legged captain. The rusty harpoons imbedded in Moby-Dick's flanks show that he has defeated lesser hunters, and by the end of the novel he has systematically destroyed Ahab's ship and crew. The 1977 film *Orca* concerns another duel between human and whale, one in which the whale has a clear motive, revenge against the hunter who murdered his mate and unborn calf. The whale is the nominal villain of the story, but his actions and motives parallel those of many human movie heroes—Mel Gibson in *Mad Max*, for example, or Clint Eastwood in *The Outlaw Josey Wales*. The aging whale-hero of Hank Searls's novel *Sounding* (1982) is driven by different, but still recognizably human, emotions. Lonely and acutely aware of his own impending death, he is drawn to a crippled Soviet submarine where Peter Rostov, the book's human hero, awaits his own inevitable death. Driven by mutual curiosity, the whale and sonar-operator Rostov form a tenuous bond that bridges the immense gap between their species and their worlds.

Scientists' sketchy understanding of whales' social and mental lives creates a vacuum that storytellers can fill in whatever ways suit the demands of their stories. We know that whales think, feel, and communicate, but not what or how. Assuming that their patterns of thought, emotion, and behavior mirror ours is, of course, an enormous convenience in telling stories about them. Indeed, it may be the *only* way to tell stories about them; science fiction writers have often suggested that a story involving a *truly* alien intelligence would, by definition, be incomprehensible. It is important though, to separate literary convention from scientific reality. We do not know, and may never know, how whale minds work. In the absence of any concrete knowledge, there is no reason to assume that they work in ways remotely similar to ours.

Related Entries: Dolphins; Intelligence, Animal

FURTHER READING

Carwardine, Mark, and Martin Camm. *DK Handbooks: Whales, Dolphins, and Porpoises*. DK Publishing, 1995. Comprehensive guide, with species-by-species illustrations.

Ellis, Richard. *Men and Whales*. Lyons Press, 1998. The history of 500 years of human-whale interaction, covering history, literature, anthropology, and folklore.

Hoyt, Erich, et al. *Whales, Dolphins, and Porpoises: A Nature Company Guide*. Time-Life Books, 1998. A compact, practical guide for lay readers; excellent illustrations.

GENERAL BIBLIOGRAPHY

The specific works consulted in the preparation of any given entry are listed in the "Further Reading and Sources Consulted" section of the entry. Some, such as Lawrence M. Krauss's invaluable *The Physics of Star Trek*, are listed under multiple entries. The works listed here are those used, throughout the book, to check the "nuts and bolts": facts, figures, dates, titles, character names, and the like. Some, like the *Internet Movie Database* and the *Facts on File Dictionaries*, are useful exclusively for such purposes. Others, like the *Encyclopedia of Science Fiction* and *Encyclopedia of Fantasy*, coedited by John Clute, also offer valuable commentary and interpretation. Works of the latter type are indicated by asterisks(*).

SCIENCE

Asimov, Isaac. *Asimov's Biographical Encyclopedia of Science and Technology*, 2nd rev. ed. Doubleday, 1982.

*Beatty, J. Kelly, et al., eds. *The New Solar System*. 4th ed. Cambridge University Press, 1998.

Brennan, Richard P. *Dictionary of Scientific Literacy*. John Wiley, 1992.

Bynum, W.E., et al. *Dictionary of the History of Science*. Princeton University Press, 1981.

Daintith, John, ed. *The Facts on File Dictionary of Chemistry*. Rev. and expanded ed., and *The Facts on File Dictionary of Physics*, rev. and expanded ed. Facts on File, 1988.

*Hazen, Robert M., and James Trefil. *Science Matters: Achieving Scientific Literacy*. Doubleday, 1991.

How Stuff Works. <http://www.howstuffworks.com>.

*Macaulay, David. *The New Way Things Work*. Rev. ed. Houghton Mifflin, 1998.

Tootill, Elizabeth, ed. *The Facts on File Dictionary of Biology*. Rev. and expanded ed. Facts on File, 1988.

POPULAR CULTURE

Altman, Mark A., and Edward Gross. *Trek Navigator: The Ultimate Guide to the Entire Star Trek Saga.* Little, Brown/Back Bay Books, 1998.

*Barron, Neil. *Anatomy of Wonder: A Critical Guide to Science Fiction.* 3rd ed. R.W. Bowker, 1987.

Brown, Charles N., and William G. Contento. *The Locus Index to Science Fiction: 1984–2001.* <http://www.locusmag.com/index/0start.html>.

*Clute, John. *Science Fiction: The Illustrated Encyclopedia.* Doring Kindersley, 1995.

*Clute, John, and John Grant. *The Encyclopedia of Fantasy.* St. Martin's Griffin, 1997.

*Clute, John, and Peter Nicholls. *The Encyclopedia of Science Fiction.* St. Martin's Griffin, 1995.

Contento, William G. *Index to Science Fiction Anthologies and Collections, Combined Edition.* <http://www.best.com/~contento/0start.html#TOC>.

The Internet Movie Database. <http://www.imdb.com>.

Maltin, Leonard. *Leonard Maltin's 1998 Movie and Video Guide.* Signet, 1997.

McNeil, Alex. *Total Television: The Comprehensive Guide to Programming from 1948 to the Present.* 4th ed. Penguin, 1996.

Rovin, Jeff. *Adventure Heroes.* Facts on File, 1994.

———. *Aliens, Robots, and Spaceships.* Facts on File, 1995.

———. *Encyclopedia of Superheroes.* Facts on File, 1985.

Videohound's Sci-Fi Experience: Your Guide to the Quantum Universe. Visible Ink Press, 1997.

Yesterdayland. <http://www.yesterdayland.com>.

Zicree, Marc Scott. *The Twilight Zone Companion.* 2nd ed. Silman-James Press, 1992.

INDEX

About the Author

A. BOWDOIN VAN RIPER is a professor in the Department of Social and International Studies at Southern Polytechnic State University. He specializes in the history of science and has written numerous articles on the history of space and aviation.